58 Advances in Polymer Science
Fortschritte der Hochpolymeren-Forschung

Polymerization Reactions

With Contributions by
R. Barbucci, P. Ferruti, E. Franta,
S. Iwatsuki, P. F. Rempp

With 21 Figures and 15 Tables

Springer-Verlag
Berlin Heidelberg GmbH
1984

ISBN 978-3-662-15243-0   ISBN 978-3-540-38738-1 (eBook)
DOI 10.1007/978-3-540-38738-1

Library of Congress Catalog Card Number 61-642

© Springer-Verlag Berlin Heidelberg 1984
Originally published by Springer-Verlag Berlin Heidelberg New York Tokyo in 1984
Softcover reprint of the hardcover 1st edition 1984

2152/3020–543210

Editors

Editorial

With the publication of Vol. 51, the editors and the publisher would like to take this opportunity to thank authors and readers for their collaboration and their efforts to meet the scientific requirements of this series. We appreciate our authors concern for the progress of Polymer Science and we also welcome the advice and critical comments of our readers.

With the publication of Vol. 51 we should also like to refer to editorial policy: *this series publishes invited, critical review articles of new developments in all areas of Polymer Science in English (authors may naturally also include works of their own).* The responsible editor, that means the editor who has invited the article, discusses the scope of the review with the author on the basis of a tentative outline which the author is asked to provide. Author and editor are responsible for the scientific quality of the contribution; the editor's name appears at the end of it.

Manuscripts must be submitted, in content, language and form satisfactory, to Springer-Verlag. Figures and formulas should be reproducible. To meet readers' wishes, the publisher adds to each volume a "volume index" which approximately characterizes the content.

Editors and publisher make all efforts to publish the manuscripts as rapidly as possible, i.e., at the maximum, six months after the submission of an accepted paper. This means that contributions from diverse areas of Polymer Science must occasionally be united in one volume. In such cases a "volume index" cannot meet all expectations, but will nevertheless provide more information than a mere volume number.

From Vol. 51 on, each volume contains a subject index.

Editors   Publisher

Table of Contents

Macromonomers:
Synthesis, Characterization and Applications

Paul F. Rempp and Emile Franta
Centre de Recherches sur les Macromolécules (CNRS), 6, rue Boussingault,
Strasbourg Cedex, France

The first part of this review critically examines the various attempts to prepare macromonomers: Anionic and cationic polymerization for synthesizing living polymers have been preferred whenever possible but two-step radical polymerizations making use of efficient transfer agents have also been applied.

In the second part, homopolymerization of the macromonomers — yielding highly branched structures — as well as copolymerization with a classical monomer — readily producing graft copolymers — are presented. The versatility of these methods is demonstrated by preparing original amphiphilic products. The properties of these segregated products are described in the few cases where they are available.

1 Introduction

This review deals with polymer species that can further participate in polymerization processes, giving thus access to graft copolymers. Macromolecular monomers, generally referred to as macromonomers or as Macromers®, are linear macromolecules carrying at their chain end some polymerizable function. In most cases, this function is an unsaturation (Fig. 1a); it can also be an oxirane ring (Fig. 1b) or another heterocycle that can undergo polymerization. Macromonomers can also be bifunctional (Fig. 1c) carrying an active double bond (or an adequately reactive heterocycle) at each end of their chain. Polymerization of such species should result in network formation. Polymer chains bearing at *one* end two functions which are able to

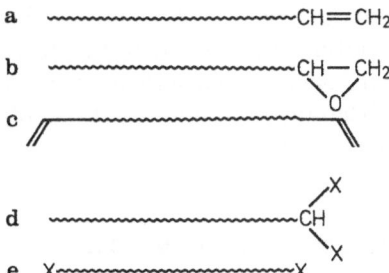

Fig. 1.

participate in a stepwise growth process (polycondensation reaction) can also be referred to as macromonomers (Fig. 1d). The latter species should not be confused with telechelic polymers (Fig. 1e) bearing two functions (either identical or antagonist) at their chain ends. Telechelic polymers can participate in further events (such as step polymerization), to yield much larger, yet linear macromolecules. Chain extension does not involve branching, and that is why such species should be referred to as telechelic polymers or α-ω functionalized polymers (thus keeping in mind that the whole chain can be incorporated into a linear polycondensate) but not as macromonomers.

The interest in macromonomers arises from the fact that they give an easy access to graft copolymers. If a vinylic or acrylic monomer is copolymerized with a macromonomer, each molecule of the latter type incorporated into the growing chain gives rise to a graft. It is usually not necessary to search for high molar percentages of macromonomer in the copolymer. However, it is necessary that both the attack of the unsaturation by an active site and the attack of the comonomer by the active site arising from the macromonomer be possible.

Macromonomers described in the literature generally exhibit rather low molecular weights (5×10^2 to 2×10^4 g · mol^{-1}). This ensures adequate characterization of the species and provides for a sufficient reaction probability of the terminal double bonds in a polymerization reaction. However, a low molecular weight is by no means a necessary limitation.

Until 1975 little work has been performed in the field of macromonomers, and most of it appeared in patent literature. Only in recent years did this domain of polymer science attract increasing interest; a large number of communications and articles, originating in many universities and industrial laboratories have been published lately, with the general aim of gaining easy access to graft copolymers. Motivations often arise from applications such as compatibilization, adhesion, coating, microemulsions, biomaterials, etc. A few short review articles have already been published on the subject [1-6].

In the first part of this review we shall consider the various pathways that have been used (or attempted) to synthesize macromolecular monomers. We shall critically discuss the efficiency of the methods that have been proposed, together with the procedures used for the characterization of the species obtained. In the second part we shall describe the various attempts to homopolymerize macromonomers and to use them in copolymerization reactions to obtain graft copolymers. We shall include some potential applications of macromonomers as intermediates to the synthesis of new polymeric materials that have been proposed.

No general method for naming macromonomers has been proposed so far, so that there is a great variety of terms used. Throughout this review we always clearly state the nature of the polymeric chain and the nature of the terminal polymerizable function. For instance, ω-methacryloyl-polytetrahydrofuran refers to a poly-THF chain bearing at the chain end a methacrylic ester function.

2 Synthesis and Characterization of Macromonomers

2.1 Historical

The first attempts to synthesize short macromolecules bearing at their chain end an active double bond were made by Greber et al. [7] in 1962. They reacted the Grignard derivative of p-chlorostyrene with ω-chlorodimethylsiloxane oligomers and obtained polydimethylsiloxane macromonomers bearing at their chain end a p.-vinyl phenyl group:

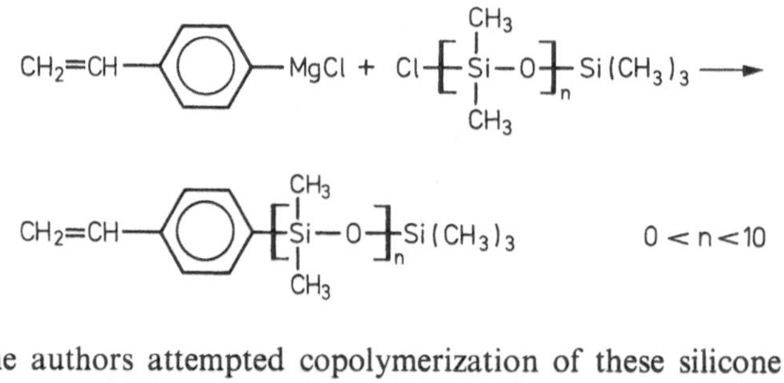

$$CH_2{=}CH{-}\underset{}{\bigcirc}{-}MgCl \ + \ Cl{-}\!\!\left[\begin{array}{c}CH_3\\|\\Si{-}O\\|\\CH_3\end{array}\right]_n\!\!\!{-}Si(CH_3)_3 \longrightarrow$$

$$CH_2{=}CH{-}\underset{}{\bigcirc}{-}\!\!\left[\begin{array}{c}CH_3\\|\\Si{-}O\\|\\CH_3\end{array}\right]_n\!\!\!{-}Si(CH_3)_3 \qquad 0<n<10$$

The same authors attempted copolymerization of these silicone — type macromonomers with different monomers (e.g. styrene) with the aim of synthesizing graft copolymers. α,ω-Diunsaturated polydimethylsiloxanes were also synthesized by the same method starting from α,ω-dichlorodimethylsiloxane oligomers.

Another early attempt to prepare low molecular weight ω-unsaturated polymers (the terms Macromer®[1] and macromonomer were not used until 1978 [6]) was made by Greber et al. [8] by means of anionic deactivation. A "living" ω-carbanionic poly-

---

[1] MACROMER® is now a trademark of CPC International Inc.

styrene was reacted with p-vinylphenyldimethyl-chlorosilane, the polystyrene chain containing then a terminal p-vinylphenyl group:

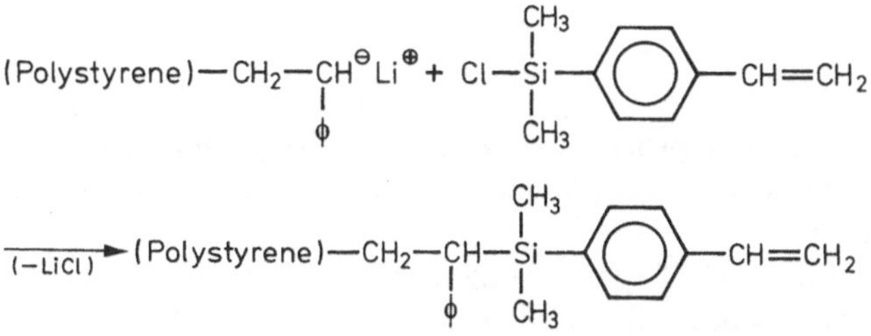

In spite of the high reactivity of the Si—Cl bond towards nucleophiles (such as styryl carbanions) it could be feared that the double bond would be attacked by the remaining carbanions. This side reaction apparently does not take place. Thus Greber showed that the number average molecular weights obtained by double bond analysis and by vapor pressure osmometry agree which each other.

A third pathway was attempted by Greber [9] who initiated the polymerization of vinyl monomers by means of the unsaturated Grignard derivative already mentioned:

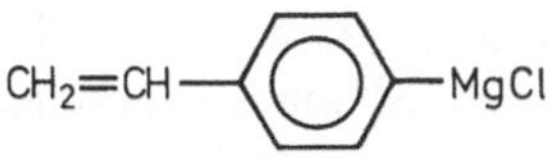

However, the efficiency of this compound as an anionic polymerization initiator is rather poor, as compared with that of standard lithiumorganic initiators.

Surprisingly, Greber did not use these ω-unsaturated polystyrenes to synthesize graft copolymers but instead to prepare block copolymers: By reacting these polystyrenes with polydimethylsiloxanes carrying — Si—H functions at both chain ends, polystyrene-b-PDMS triblock copolymers are obtained. By metalation of the terminal unsaturation with alkali metals followed by addition of vinylpyridine Greber synthesized polystyrene-polyvinylpyridine block copolymers; these molecules should exhibit a branched structure (owing to radical recombination prior to the anionic initiation of the vinylpyridine) but this is not discussed in Greber's papers.

Another early attempt to synthesize macromonomers was performed by Gillman and Senogles [10] on a quite different basis but with the final aim of preparing graft copolymers. The free-radical polymerization of methyl methacrylate was carried out in the presence of 2-hydroxyethanethiol ($HO-CH_2-CH_2-SH$). This compound is an efficient transfer agent, thus lowering considerably the molecular weight of the polymer (with respect to its value in the absence of the thiol), and the polymer molecules are fitted with $-S-CH_2-CH_2OH$ end groups originating from the transfer reaction. Next, Gillman and Senogles reacted the ω-hydroxy polymer with metharyloyl chloride thus introducing into this polymer terminal methacrylic ester functions:

$$\text{Poly (methyl methacrylate)} - \underset{\underset{\text{COOCH}_3}{|}}{\overset{\overset{\text{CH}_3}{|}}{C}} - \text{CH}_2 - S - \text{CH}_2 - \text{CH}_2 - O - CO - \underset{\text{CH}_3}{\overset{\text{CH}_2}{C}}$$

The reaction was shown to proceed nearly quantitatively, and the poly(methyl methacrylate)macromonomers were copolymerized with various vinylic monomers to yield graft copolymers.

A more general method of synthesizing macromonomers by means of free-radical transfer processes has been disclosed in a Dutch patent of ICI [11] dated 1965. A number of examples were given but precise data on the species formed are missing.

These early examples of macromonomer synthesis illustrate some of the chief principles that have subsequently led to a great variety of methods. Ionic polymerization is often preferred because of the long lifetime of the active sites. Transfer reactions in free-radical processes are also used quite often, yielding both acceptable molecular weights and an adequate proportion of terminal functions.

We shall distinguish between the methods of synthesis that give "direct" access to the macromonomer (i.e. in one step) and those involving two steps. In the two-step method functionalization at the chain end is performed followed by reaction with some unsaturated compound to introduce into the macromolecule a polymerizable double bond at chain end.

Adequate characterization of the species formed is essential, as this is a control of the ability of a given method to yield well-defined macromonomers. We shall therefore emphasize the methods of characterization that have been chosen in the various cases.

2.2 Macromonomer Synthesis Using Anionic Polymerization Methods

Anionic polymerizations, when carried out in aprotic solvents, are characterized by the long lifetime of the carbanionic (or oxanionic) sites [12]. When neither spontaneous transfer nor termination reactions are involved, the polymers obtained exhibit sharp molecular weight distributions, and their number average degree of polymerization is determined by the [Monomer]/[Initiator] molar ratio, provided initiation is fast as compared to propagation. However, the major advantage of these methods, as far as synthesis is concerned, is the socalled "living" character of the polymers [12]: After completion of the polymerization the active sites retain their reactivity and can be used for functionalizations at the chain end.

Two *direct* pathways can be imagined, to fit a polymer molecule at its chain end with a polymerizable unsaturation.

— In the first method an unsaturated metalorganic initiator is used. If the initiation proceeds by addition to the monomer chosen (and not by electron transfer) and if the active double bond of the initiator is insensitive to the attack by carbanions (or oxanions) the process should yield a macromonomer.

— Alternately, use can be made of the reactivity of the active site at the chain end after the polymerization reaction (initiated by a standard metalorganic compound) has gone to completion. An unsaturated electrophile added at this point should deactivate the sites, thus introducing a bond between the polymer chain and the unsaturation, provided no side reaction affects the latter.

Anionic polymerization has also been used as a route to various *indirect* macromonomer syntheses. Functionalizations are easy to carry out anionically, and the terminal functions thus created can be reacted further, e.g. with unsaturated derivatives.

All pathways involving anionic polymerization allow in principle an adequate control of the molecular weight and ensure low polydispersities. They often lead to well defined macromonomers. We shall now present the various procedures published in articles and patents and critically appraise the results obtained and the limitations of the methods proposed.

2.2.1 Anionic Initiation Procedures

Waak [13-15] was the first to attempt anionic initiation by means of an unsaturated lithiumorganic compound, namely vinyllithium. This reagent can be obtained by an exchange reaction between tetravinylzinc and butyllithium:

$$(CH_2{=}CH)_4 \, Zn \; + \; 4 \, BuLi \longrightarrow 4 \, CH_2{=}CHLi \; + \; Bu_4Zn$$

Vinyllithium initiates the polymerization of styrene and of similar monomers. However, the rate of initiation is fairly low, and the molecular weights exhibited by the polymers obtained are by one order of magnitude higher than the expected ones. This is not the only drawback of the method. A macromolecule obtained by anionic polymerization using vinyllithium carries an allylic double bond at the chain end, and this type of unsaturation is not adequate for the subsequent radical copolymerization because of its low reactivity towards radicals and of isomerization that can take place.

$$CH_2{=}CH{-}CH_2{-}\underset{\underset{\phi}{|}}{CH}{-}CH_2{-}\underset{\underset{\phi}{|}}{CH}\ldots\ldots$$

Waak [15] also used other unsaturated lithium organic initiators (such as allyllithium, crotyllithium) for styrene polymerization. Though more efficient than vinyllithium, these initiators exhibit the same disadvantages. Therefore, the same conclusions apply.

The use of Grignard compounds derived from p-chlorostyrene has already been mentioned as one of the early attempts to synthesize macromonomers [9]. Vinylpyridine and methyl methacrylate were polymerized. However, no precise data were given on the efficiency of this unsaturated Grignard initiator nor on the polydispersity of the functional polymers obtained; the possibility of side reactions (involving, for instance, the ester function of methyl methacrylate) was not discussed either.

However, the initiation by means of an efficient unsaturated initiator is impeded by the fact that the unsaturation is generally sensitive to the attack by carbanions whereby branched structures are formed.

Quite different is the case of polymerizations involving oxanionic sites since alkoxides do generally not react with carbon-carbon double bonds. The ring-opening polymerization of oxirane can be initiated by alkoxides, and the synthesis of polyoxyethylene macromonomers was achieved by means of an unsaturated alcoholate [16]. Potassium p-isopropenyl benzylate has proved to be an efficient initiator for the polymerization of oxirane causing no side reactions. Thus, the poly(ethylene oxides) obtained contain at their chain end an α-methylstyryl residue originating from the initiator. When the polymerization is completed the "living" ends are deactivated either by protonation or by means of an alkyl halide:

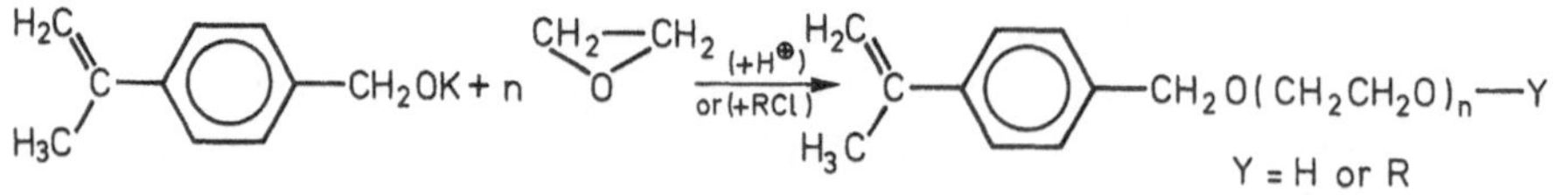

The preparation of the potassium alcoholate requires much care. Metalation of the alcohol is carried out to completion without attack of the double bond by means of diphenylmethyl potassium in THF solution. Excess alcohol present in the subsequent polymerization system would act as a transfer agent.

The reaction medium is heterogeneous at the onset of the polymerization because the potassium alcoholates associate and are insoluble in THF. Only after the chains have reached an average polymerization degree of the order of 5 does it become homogeneous again. The heterogeneous initiation has no practical consequences for the width of the molecular weight distribution of the macromonomer, even for M_n values of the order of 1000. The covered molecular weight range extends from 800 to 10000, and a very extensive characterization of the obtained samples (by light scattering, NMR, GPC, double bond analysis, UV spectrometry) has revealed that the reactions take place quantitatively and that the macromonomers exhibit the expected structure.

A very recent paper by Saegusa [17] describes a method of synthesizing polyoxyethylene macromonomers bearing a polymerizable heterocycle at the chain end. Here again the method involves initiation of the oxirane polymerization by means of an alcoholate (derived from 2-p-hydroxyphenyl)-oxazoline). As metalation agent butyllithium, was used since lithium alcoholates are not very reactive towards oxirane. Indeed, the macromonomers obtained exhibit very low degrees of polymerization. Deactivation was performed with methyl iodide:

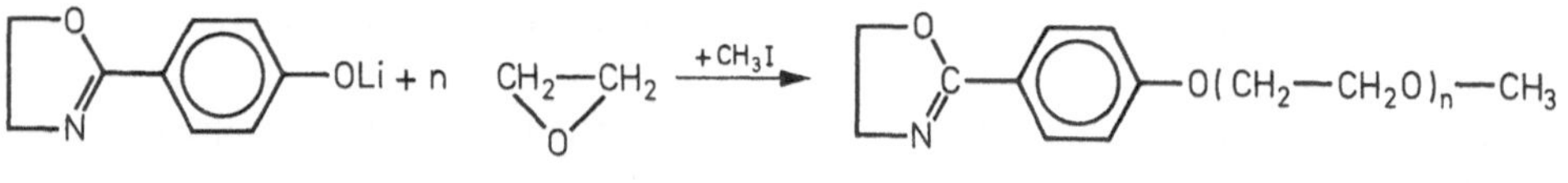

This species can undergo cationic homopolymerization; cationic copolymerization with another heterocyclic monomer has also been attempted.

2.2.2 End-Capping of Living Polymeric Anions

A great deal of work most of which was published in patents was carried out along this line. As mentioned already, the basic principle of anionic deactivation is to react a monofunctional "living" polymer with an unsaturated electrophile. It must be ensured that this deactivation predominates with respect to the attack of the active double bond by carbanions. The electrophiles that are most commonly used include organic halides and esters. It should however be taken into account that deactivation of carbanions by means of electrophiles (Wurtz — type reactions) is not always fast and quantitative and that undesirable side reactions may occur. A proper characterization of the species obtained is therefore always essential.

Unsaturated electrophiles that have been employed for the purpose of synthesizing macronomers are compiled in Table 1.

— *Allyl halides* [18, 19] are efficient deactivators for "living" polymers although the reaction is not always quantitative and free of side reactions. In the case of polystyrene allyl groups are introduced quantitatively as end groups into the macromolecules

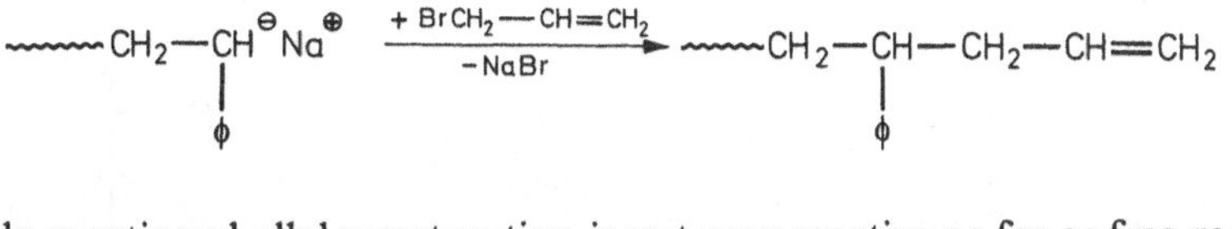

As already mentioned allyl unsaturation is not very reactive as far as free-radical polymerizations are concerned

In a recent paper Sharkey et al. [20] have shown that carbanions of lower reactivity such as those originating from methyl methacrylate do react with organic halides such as allyl bromide. The reaction proceeds quantitatively and rapidly at — 78 °C in THF solution. No side reaction has been detected.

— *Benzyl halides* are known to be efficient deactivators for living polystyrene like allyl halides. It can be expected, however, that the reaction of a styryl carbanion with p-vinylbenzyl chloride competes with a side reaction involving attack of the carbanion at the double bond of p-vinylbenzyl chloride (VBC):

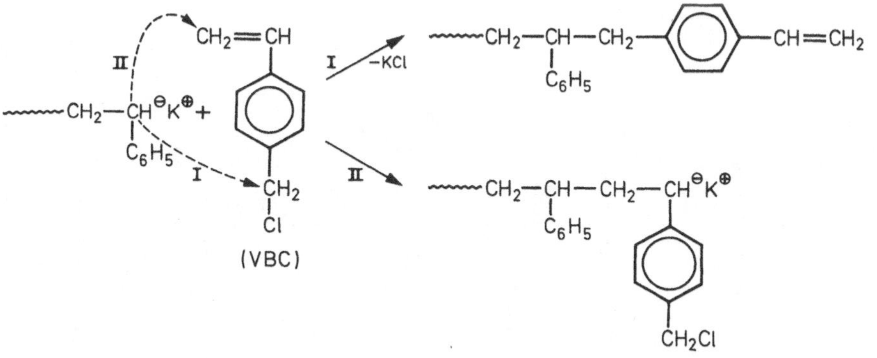

Asami [21] studied this reaction. Even in the presence of a tenfold excess of VBC, is the reaction strongly influenced by the nature of the solvent, lithium being the counterion. Asami established that when the reaction medium contains tetrahydro-furan (THF) no side reaction (leading *in fine* to a molecular weight increase) was detected. The polystyrene obtained contains at its chain end a styryl group.

There are other possibilities of suppressing side reaction II. Thus the addition of 1,1-diphenylethylene [18, 22] or oxirane [18, 23] causes a decrease of the nucleophilicity of the "living" sites prior to their reaction with VBC. This was especially tested by

Table 1.

Initiator	Monomer	Deactivator	Product	Ref.
CuK	Styrene	$BrCH_2-CH=CH_2$	$\sim\sim CH_2-CH(C_6H_5)-CH_2-CH=CH_2$	18,19
BuLi	Styrene	$BrCH_2-CH=CH_2$	$\sim\sim CH_2-CH(C_6H_5)-CH_2-CH=CH_2$	18,19
BuLi	MMA	$BrCH_2-CH=CH_2$	$\sim\sim CH_2-C(CH_3)(COO-CH_3)-CH_2-CH=CH_2$	20
BuLi or	Styrene	$Cl-CO-C(=CH_2)CH_3$	$\sim\sim CH_2-CH(C_6H_5)-CO-C(=CH_2)CH_3$	18
CuK	Styrene	CH_2-CH_2(oxirane) $/ClCO-C(=CH_2)CH_3$	$\sim\sim CH_2-CH(C_6H_5)-CH_2-CH_2-O-CO-C(=CH_2)CH_3$	18,24
$(C_6H_5)_2CHK$ or ROK	Oxirane	$Cl-CO-C(=CH_2)CH_3$	$\sim\sim CH_2-CH_2-O-CO-C(=CH_2)CH_3$	16
CuK/BuLi	Styrene	$Cl-CH_2-Si(CH_3)(CH_3)-C(=CH_2)CH_3$	$\sim\sim CH_2-CH(C_6H_5)-CH_2-Si(CH_3)(CH_3)-C=CH_2$	28
CuK	Styrene	$Cl-Si(CH_3)(CH_3)-HC=CH_2$	$\sim\sim CH_2-CH(C_6H_5 CH_3)-Si(CH_3)-CH=CH_2$	28

Initiator	Monomer	Deactivator	Product	Ref.
BuLi or	Styrene	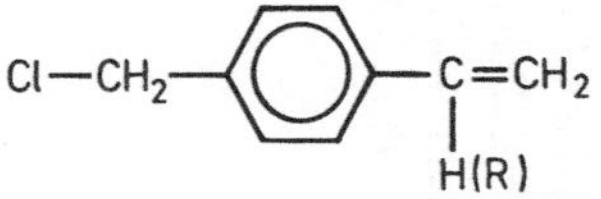 $Cl-CH_2-$ (with $C=CH_2$, $H(R)$)	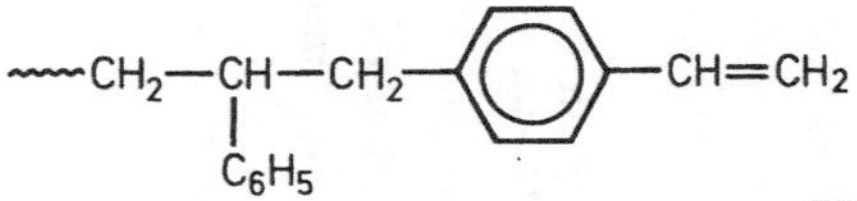 $\sim\sim CH_2-CH-CH_2-$ (with C_6H_5) phenyl ring $-CH=CH_2$	21)
CuK	Styrene	CH_2-CH_2 / $ClCH_2$ (epoxide, O) ring $-CH=CH_2$	$\sim\sim CH_2-CH-CH_2-CH_2-O-CH_2-$ (with C_6H_5) ring $-CH=CH_2$	18)
	MMA	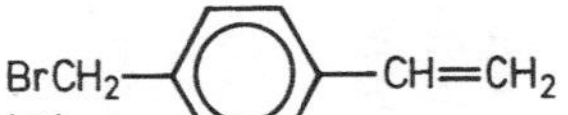 $BrCH_2-$ ring $-CH=CH_2$ (I)	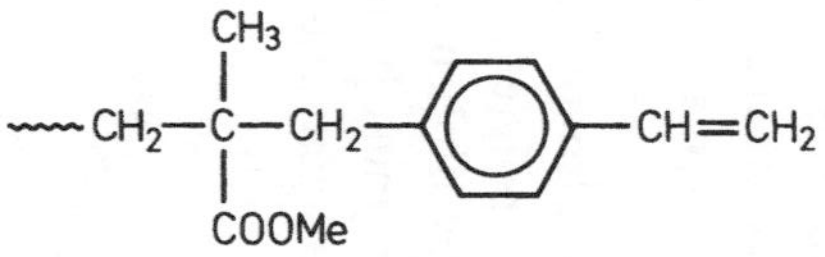 $\sim CH_2-C(CH_3)-$ (with COOMe) ring $-CH=CH_2$	20)
$(CH_3)_3Si-$	D_3	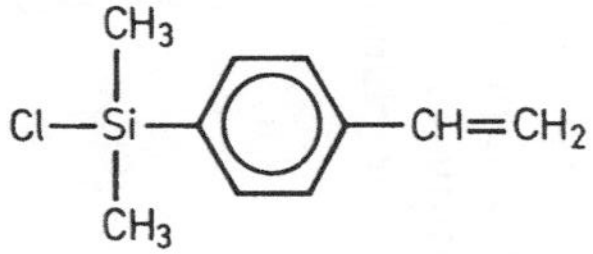 $Cl-Si(CH_3)(CH_3)-$ ring $-CH=CH_2$	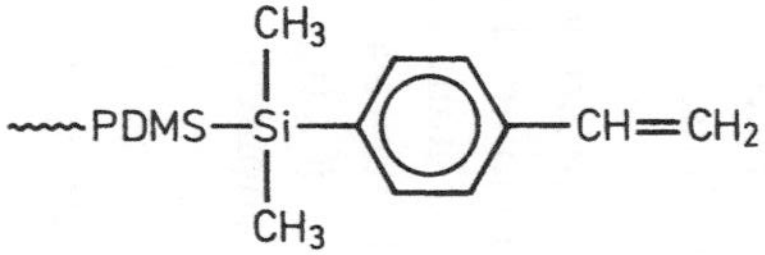 $\sim\sim PDMS-Si(CH_3)(CH_3)-$ ring $-CH=CH_2$	29)
OLi		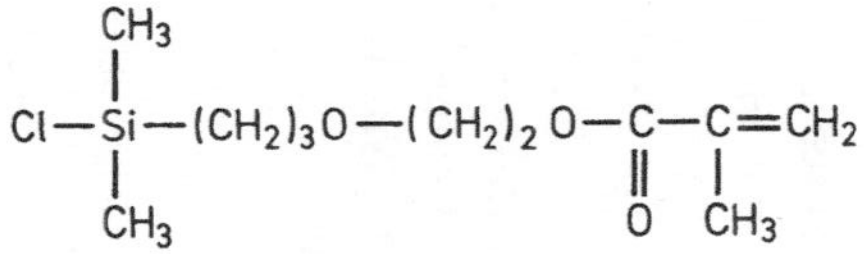 $Cl-Si(CH_3)(CH_3)-(CH_2)_3O-(CH_2)_2O-C(=O)-C(CH_3)=CH_2$	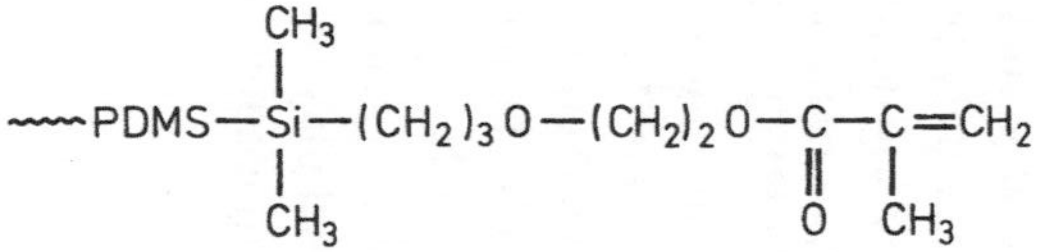 $\sim\sim PDMS-Si(CH_3)(CH_3)-(CH_2)_3O-(CH_2)_2O-C(=O)-C(CH_3)=CH_2$	29)

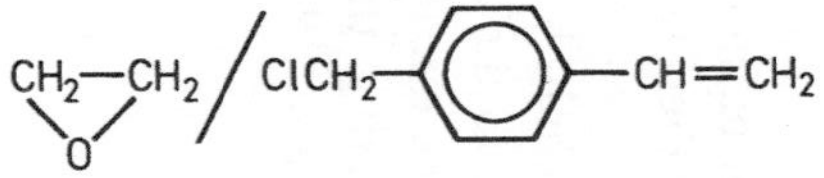

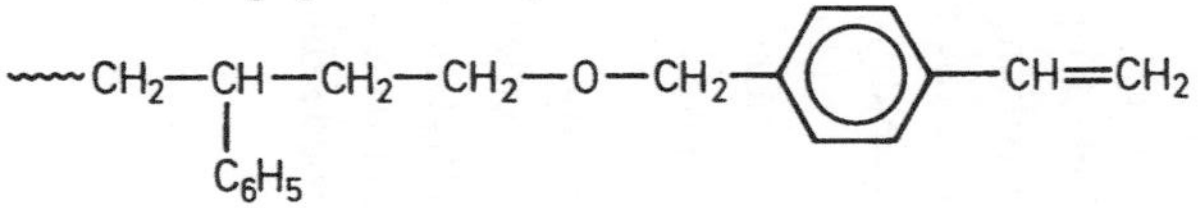

Milkovich [18], and was found to be a satisfactory means to get rid of side reaction II.

Sharkey [20] found that "living" poly(methyl methacrylate) reacts efficiently and without side reaction with p-vinylbenzyl iodide (or bromide) at low temperature yielding poly(methyl methacrylate) macromonomers bearing at the chain end a styryl group (ω-styryl-poly(methyl methacrylate)macromonomer).

— *Methacryloyl chloride* has also been tested as a potential electrophilic deactivator for "living" polystyrenes [18]. However, side reactions involving the methacrylic double bond can only be avoided if the nucleophilicity of the styryl carbanions is decreased by the addition of 1,1-diphenylethylene (yielding diphenylmethyl carbanions [22]) or oxirane (yielding alkoxide sites [23]) prior to the addition of the acid chloride [18,24]. The latter method has proved satisfactory and yields polystyrene macromonomers of well defined molecular weights and narrow molecular weight distributions [24].

Poly(ethylene oxide)macromonomers were also prepared by reaction of methacryloyl chloride with "living" monofunctional poly(ethylene oxide) [16] the polymerization of which can be efficiently initiated by diphenylmethyl potassium [25] or by potassium alkoxides [26]. In the latter case care must be taken to avoid any excess of alcohol which would play the role of a transfer agent and impede the macromonomer synthesis. The obtained macromonomers

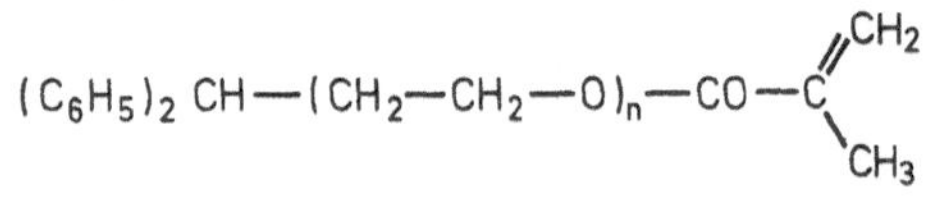

$$(C_6H_5)_2\, CH-(CH_2-CH_2-O)_n-CO-C\underset{CH_3}{\overset{CH_2}{\lessgtr}}$$

can also be synthesized in two steps. The polymerization of oxirane is followed by protonation, and the terminal hydroxy group can be further reacted as shown below.

The introduction of vinylsilyl end groups into a polymer chain has been achieved by several techniques. Since vinylsilyl functions are sensitive to the attack by strong nucleophiles [27], it was considered appropriate to decrease the nucleophilicity of the styryl carbanions by the addition of oxirane before reacting them with either chloromethyldimethylvinylsilane or chlorodimethylvinyl silane [28a]. The following structures are formed:

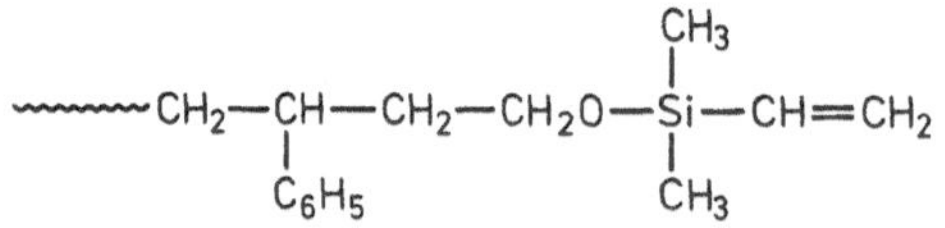

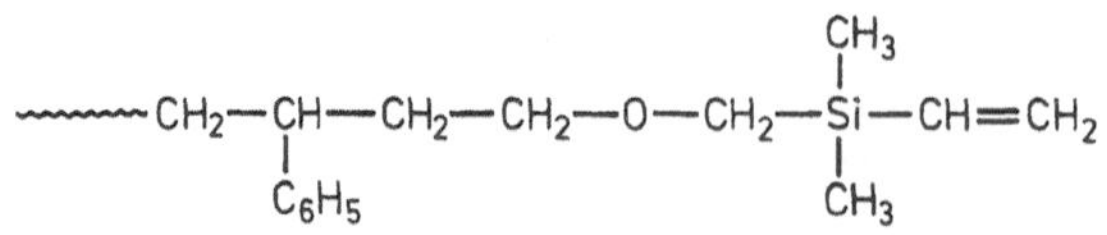

$$\sim\!\!\sim\!\!\sim CH_2-\underset{\underset{C_6H_5}{|}}{CH}-CH_2-CH_2O-\underset{\underset{CH_3}{|}}{\overset{\overset{CH_3}{|}}{Si}}-CH\!=\!CH_2$$

$$\sim\!\!\sim\!\!\sim CH_2-\underset{\underset{C_6H_5}{|}}{CH}-CH_2-CH_2-O-CH_2-\underset{\underset{CH_3}{|}}{\overset{\overset{CH_3}{|}}{Si}}-CH\!=\!CH_2$$

These structures have been obtained in high yields, but they are somewhat sensitive to hydrolysis. It was found [28a] that, provided it is carried out at very low temperature ($< -75\,°C$), the "direct" reaction of styryl carbanions with chlorodimethylvinylsilane yields excellent results, and the macromonomers thus formed are quite stable:

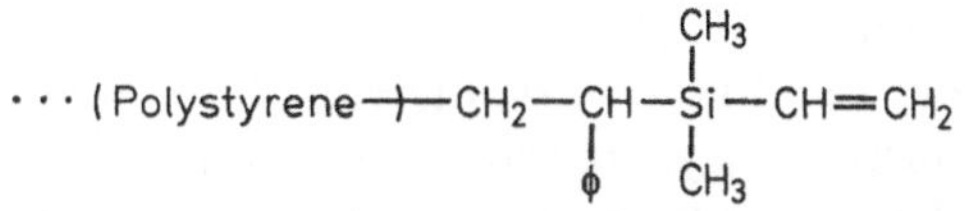

$$\cdots \text{(Polystyrene} \rightarrow \text{—CH}_2\text{—CH—Si—CH=CH}_2$$

The ability of these macromonomers to participate in free-radical copolymerization was not tested. They were used [28b] for the synthesis of tri- and multiblock copolymers by means of hydrosilylation, i.e. an addition reaction between vinylsilyl groups and the —Si — H functions at the chain end of poly(dimethylsiloxane)molecules.

A number of other unsaturated electrophilic compounds were used by Milkovich [18] as deactivators for living polystyrene or living polydienes. A characterization of the macromonomers obtained showed that the reaction of the living polymer with compounds such as maleic anhydride, vinyl chloroacetate, or 2-chloroethylvinyl ether yields the following unsaturated chain ends (in some cases the addition of 1,1-diphenylethylene is necessary):

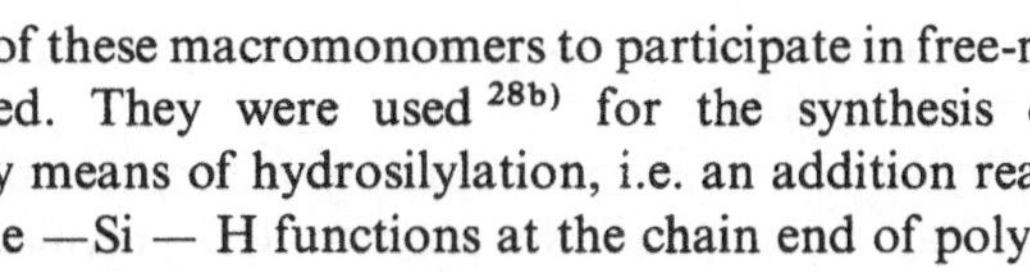

$$\text{⁓CH}_2\text{—CH—CO—CH=CH—COOH}$$
$$\overset{|}{\text{C}_6\text{H}_5}$$

$$\text{⁓CH}_2\text{—CH—CH}_2\text{—C—CH}_2\text{—CO—O—CH=CH}_2$$

$$\text{⁓CH}_2\text{—CH—CH}_2\text{—CH}_2\text{—O—CH=CH}_2$$
$$\overset{|}{\text{C}_6\text{H}_5}$$

Milkovich et al. [18] reacted living polystyrene with epichlorohydrin and, after hydrolysis of the epoxide ring, obtained quantitatively a macromonomer possessing at one chain end two adjacent hydroxy groups which can subsequently add to diisocyanates to yield "graft polycondensates"

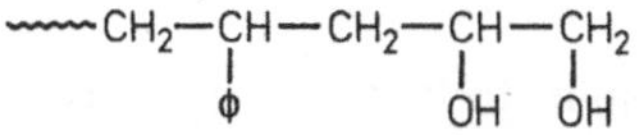

$$\text{⁓CH}_2\text{—CH—CH}_2\text{—CH—CH}_2$$

The synthesis of poly(dimethylsiloxane)macromonomers was also carried out anionically. The cyclic trimer D_3 of dimethylsiloxane may be polymerized anionically, using lithium trimethylsilanolate as the initiator. Deactivation of the silanolate end groups may be performed with chlorosilyl functions. Yamashita et al. [29] successfully applied the unsaturated deactivators I and II

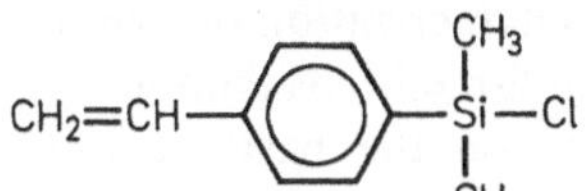

$$\text{CH}_2\text{=CH—}\bigcirc\text{—Si—Cl} \qquad \text{CH}_2\text{=C—C—O—CH}_2\text{—CH}_2\text{—O—CH}_2\text{—CH}_2\text{—CH}_2\text{—SiCl}$$

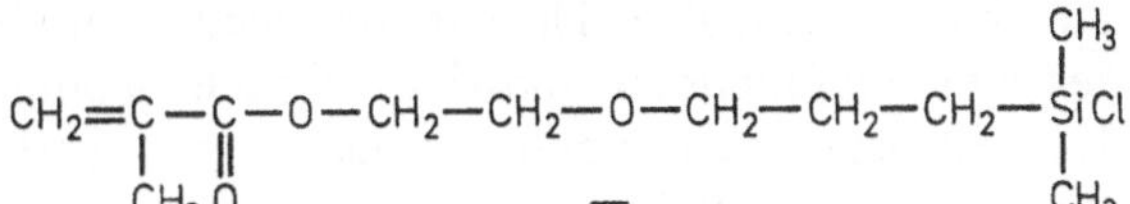

and obtained the expected macromonomers in good yields. The molecular weight range covered extended from 3000 to 8000 and the agreement between the experimentally determined molecular weights (vapor pressure osmometry, UV spectroscopy, and GPC) and the theoretical values was good. The polydispersity index was of the order of 1.1. The deactivators I and II were synthesized, from p-chlorostyrene and from 2-allyloxyethanol and methacryloyl chloride, respectively.

Thus, Yamashita succeeded in synthesizing ω-methacryloyl and ω-styryl-poly-dimethylsiloxane macromonomers which have subsequently been copolymerized with styrene and methyl methacrylate.

Difunctional polydimethylsiloxane macromonomers were prepared by Katz et al. [30] in a completely different way using a cationic process. The polymerization of the cyclic tetramer of poly(dimethylsiloxane) (D_4) was carried out in the presence of a disiloxane containing carboxy groups. Since this compound acts as a transfer agent carboxylic groups are introduced as end groups into the polymer molecules. These groups are subsequently reacted with 2-hydroxyethyl acrylate. The obtained diunsaturated PDMS macromonomers

$$
\begin{array}{ccc}
 & CH_3 & CH_3 \\
 & | & | \\
HOOC-(CH_2)_2-Si-O-Si-(CH_2)_2\,COOH \\
 & | & | \\
 & CH_3 & CH_3
\end{array}
$$

may be photopolymerized (or copolymerized with styrene) to yield the PDMS networks.

2.2.3 "Indirect" Methods using End-Capping of Living Polymeric Anions

Anionic deactivation processes can also be used to introduce quantitatively into a polymer chain terminal functions which can be utilized for the synthesis of macro-monomers.

Styryl carbanions readily react with carbon dioxide [31] to yield carboxy end groups. These terminal groups are also introduced by reaction with anhydrides [19] whereas the use of oxirane [23] leads to the formation of hydroxy end groups; esters and nitriles are used to introduce carbonyl functions at the chain end. These reactions can be carried out at low temperature in THF solution and proceed quantitatively if no deactivation by protons occurs.

Diene carbanions exhibit a reactivity similar to that of styryl carbanions. Carban-ionic sites of lower reactivity can be functionalized under certain conditions [32]. Living polymers with alkoxide end groups exhibit the same reactivity as alcoholates with respect to proton-donating substances and activated organic halides. Protonation yields terminal hydroxy groups which are often used in the two-step macromonomer synthesis.

Poly(ethylene oxide)macromonomers have thus been synthesized by two-step processes. Poly(oxyethylene)monomethyl ethers with widely varying molecular weights are commercially available. They are obtained by anionic polymerization of oxirane initiated by monofunctional alkoxides [26] such as potassium 2-methoxyethanol.

Polyoxyethylene macromonomers have been synthesized on the basis of PEO chains bearing at *one* end a terminal hydroxy group.

Thus, Graetz [33] reacted functional PEO prepolymer first with a cyclic anhydride and subsequently with glycidyl methacrylate. Efficient removal of the water in the monohydroxy PEO is however required before starting the synthesis. This is achieved by azeotropic toluene distillation:

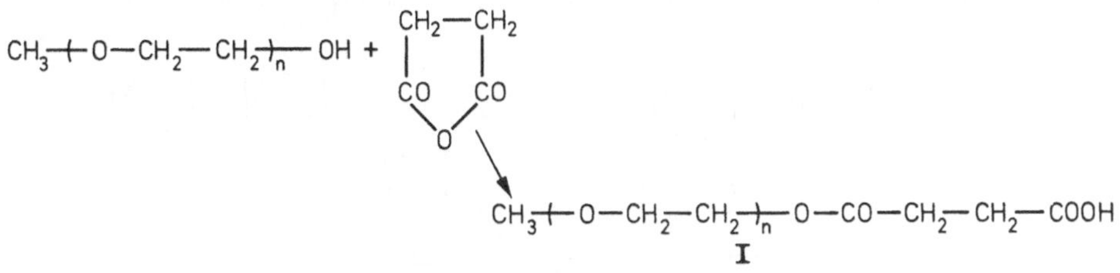

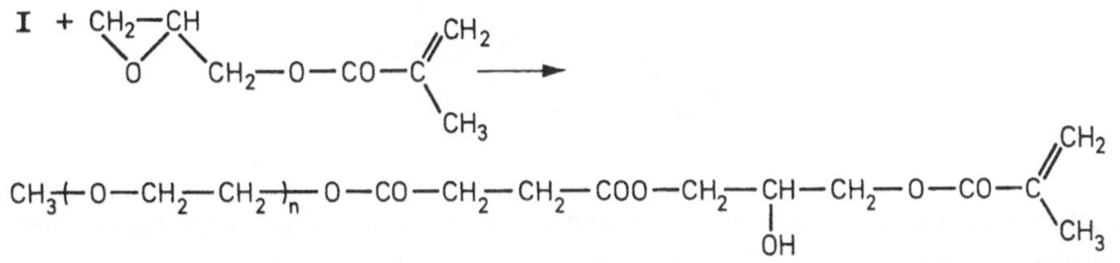

$$CH_3 \leftarrow O-CH_2-CH_2 \xrightarrow{}_n OH \; + \; \begin{array}{c} CH_2-CH_2 \\ | \qquad | \\ CO \quad CO \\ \diagdown O \diagup \end{array}$$

$$CH_3 \leftarrow O-CH_2-CH_2 \xrightarrow{}_n O-CO-CH_2-CH_2-COOH$$
$$\text{I}$$

$$\text{I} \; + \; \begin{array}{c} CH_2-CH \\ \diagdown O \diagup \quad CH_2-O-CO-C \diagup^{CH_2} _{\diagdown CH_3} \end{array} \longrightarrow$$

$$CH_3 \leftarrow O-CH_2-CH_2 \xrightarrow{}_n O-CO-CH_2-CH_2-COO-CH_2-\underset{\underset{OH}{|}}{CH}-CH_2-O-CO-C\diagup^{CH_2}_{\diagdown CH_3}$$

In another method Gramain and Frere [34] reacted Polyoxyethylene monomethyl ether with methacryloyl chloride under various experimental conditions. The reaction was carried out in the presence of a tertiary amine and yielded the expected macromonomers. Care was taken to avoid two possible side reactions, namely addition of HCl to the double bond, and polymerization of the methacrylic acid derivatives. Applying appropriate reaction conditions, the authors obtained ω-methacryloyl PEO macromonomers quantitatively.

A third method, first introduced by Stowe [35] and studied in greater detail by Guyot et al. [36], consists in reacting a polyoxyethylene monomethyl ether in alkaline medium with p-chloromethylstyrene (p-vinylbenzyl chloride, VBC). The macromonomers containing terminal p-vinylbenzyl groups are formed in good yields and have been adequately characterized.

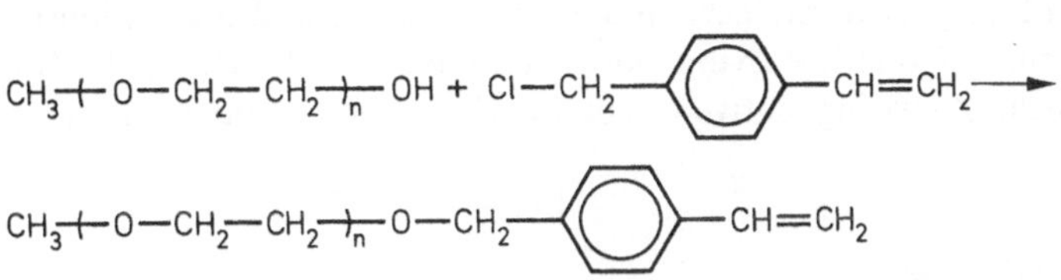

$$CH_3 \leftarrow O-CH_2-CH_2 \xrightarrow{}_n OH \; + \; Cl-CH_2-\underset{}{\bigcirc}-CH=CH_2 \longrightarrow$$

$$CH_3 \leftarrow O-CH_2-CH_2 \xrightarrow{}_n O-CH_2-\underset{}{\bigcirc}-CH=CH_2$$

These macromonomers have been copolymerized with various comonomers.

2.2.4 Double Bond Formation by Spontaneous Deactivation

When living polystyrene anion is kept for very long periods or heated to temperatures around 60 °C, a color change occurs. This is due to the elimination of sodium hydride from the active sites leading to the formation of a double bond [37]. However, this reaction proceeds slowly and cannot be used for the synthesis of macromonomers because the species formed is metalated

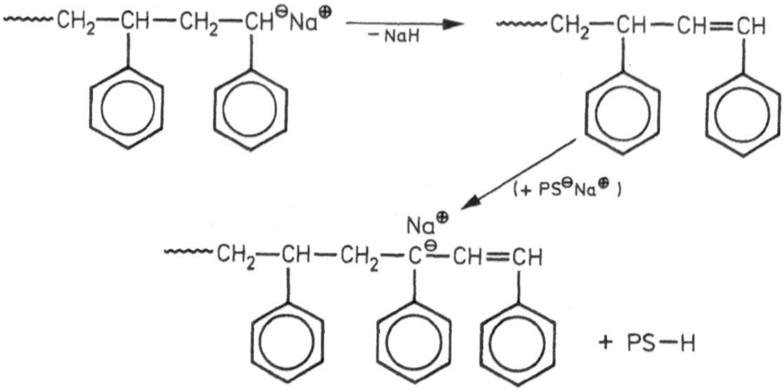

immediately by exchange reaction with a regular styryl anion. Thus, only half of the macromolecules bear terminal double bonds, the others are protonated.

Recently, the stability of "living" poly-p-bromostyrene was investigated by Jagur [38] who postulated a radical mechanism to account for the observed phenomena. Here again terminal double bonds are formed, KBr being eliminated:

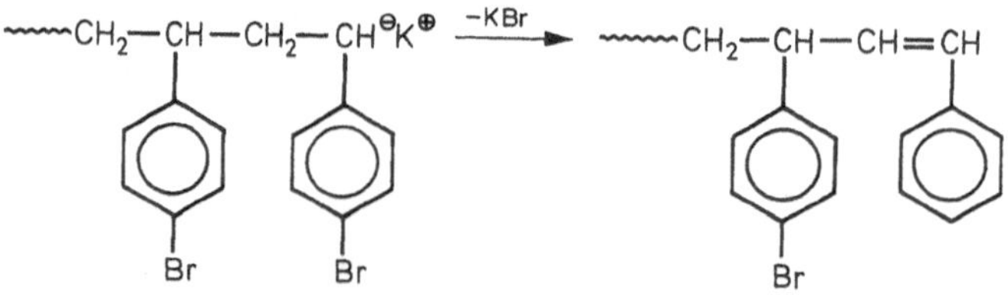

However, as in the preceding case, the hydrogen (both allylic and benzylic) in β-position with respect to the carbanionic site is acidic and metalation occurs readily at the expense of another living polymer molecule. Therefore, the fraction of the macromolecules bearing active carbon-carbon double bonds at a chain end is $\simeq 0.5$.

2.2.5 Other Process

Recently, a very unique method was reported [136] which might be amenable to some general applications and therefore worth mentioning here.

In this method mono- or dihydroxy-terminated polyoxyethylene oligomers are reacted in bulk with potassium metal at 120 °C, thus converting quantitatively the hydroxy groups into potassium alcoholate functions. Then, the medium is heated up to 180 °C and acetylene is added. This addition to the alcoholates quantitatively produces mono- or divinyl-terminated polyoxyethylenes, respectively:

$$HO(CH_2CH_2O)_n\!-\!CH_3 \xrightarrow[\substack{2)\,CH\equiv CH \\ (180\,°C)}]{1)\,K(N_2)} CH_2 = CH\!-\!O(CH_2CH_2O)_n\,CH_3$$

Since some chain shortening takes place during the process the different compounds have to be separated by fractional distillation.

Homopolymerization and copolymerization with ethylvinyl ether were attempted but no characterization of the obtained polymers has been reported so far. The formed compounds were used as polymeric catalysts in interfacial reactions and showed a significant activity.

2.3 Macromonomer Synthesis Using Cationic Methods

We shall discuss separately the polymerization of heterocyclic monomers and vinylic monomers since the mechanism of the cationic polymerization is quite different for these two types of monomers. In contrast to some heterocyclic monomers olefinic and vinylic monomers do not produce living cationic species. Therefore, the methods for synthesizing macromonomers differ greatly.

2.3.1 Heterocyclic Monomers

The cationic ring-opening polymerization of some heterocycles such as oxolane (tetra-hydrofurane, THF) with some specific initiators proceeds free of spontaneous termination reactions and without transfer to monomer [39,40]. In other words, the lifetime of the active sites is long and the polymers are "living". A special feature of this type of systems has to be taken into account: Propagation is a reversible process [39] and at each temperature an equilibrium monomer concentration can be defined. It should be remembered that the polydispersity tends to broaden if conditions approaching the "ceiling temperature" are used [41].

To initiate efficiently, rapidly and quantitatively the polymerization of these heterocycles it was searched for cationic species which add to the monomer and are associated to stable counterions. Suitable initiators are oxocarbenium salts (I), stabilized carbenium salts (II) and trifluorosulfonic derivatives (III) (triflic esters or anhydrides) [42−44]:

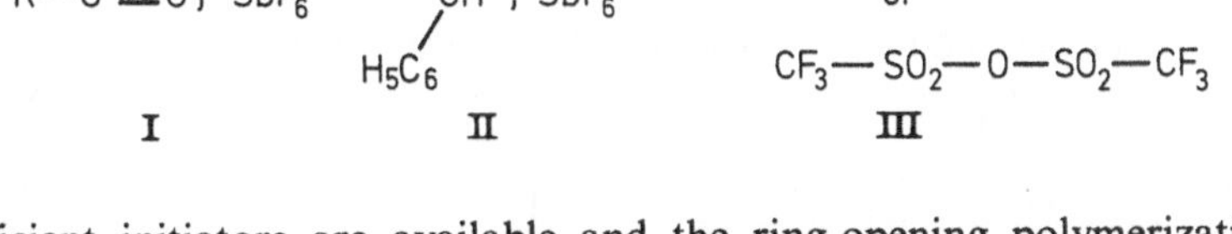

$$R\!-\!\overset{\oplus}{C}\!\!\equiv\!\!O,\ SbF_6^{\ominus} \qquad\qquad \underset{H_5C_6}{\overset{H_5C_6}{\diagdown}}\!\!\overset{\oplus}{C}H,\ SbF_6^{\ominus} \qquad\qquad \begin{array}{c} CF_3\!-\!SO_3\!-\!CH_3 \\ or \\ CF_3\!-\!SO_2\!-\!O\!-\!SO_2\!-\!CF_3 \end{array}$$

$$\qquad\quad \textbf{I} \qquad\qquad\qquad\qquad \textbf{II} \qquad\qquad\qquad\qquad \textbf{III}$$

Since efficient initiators are available and the ring-opening polymerization of oxolane yields "living" species, the basic principles of macromonomer synthesis that have been developed above should also apply to the following case:

— An unsaturated cationic initiator can be used if its double bond cannot take part in polymerization and if initiation proceeds exclusively by addition (and neither by hydride shift nor by proton transfer).

— Alternatively, a living polymer chain can be deactivated by reacting its active site with an unsaturated nucleophile. Here again the double bond should not undergo side reactions.

Both methods have been used for the synthesis of poly-THF macromonomers.

2.3.1.1 Cationic Initiation

As already mentioned, oxocarbenium salts are efficient initiators [42,45] for the polymerization of oxolane. They can be prepared *in situ* by a metathetic reaction between the corresponding acyl halide and silver hexafluoroantimonate:

$$R-\overset{O}{\underset{Cl}{C}} + AgSbF_6 \longrightarrow R-CO^{\oplus}, SbF_6^{\ominus} + AgCl\downarrow$$

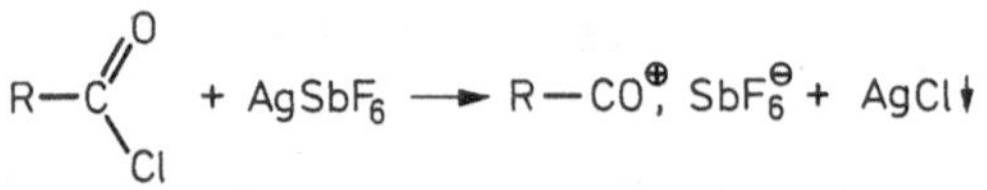

These salts are relatively stable, but if the reaction is carried out in oxolane, initiation occurs immediately:

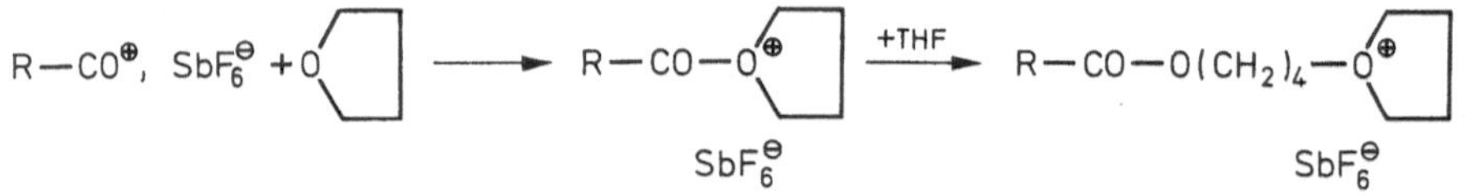

$$R-CO^{\oplus}, SbF_6^{\ominus} + O\langle\rangle \longrightarrow R-CO-O\overset{\oplus}{\langle\rangle} \underset{SbF_6^{\ominus}}{\xrightarrow{+THF}} R-CO-O(CH_2)_4-O\overset{\oplus}{\langle\rangle}$$

Methacryloyl chloride was used for the preparation of an unsaturated cationic initiator [46]. Since the methacrylic carbon-carbon double bond does not polymerize cationically it is not involved in this process. Kinetic measurements showed that methacryloyl hexafluoroantimonate is an efficient initiator, which reacts quantitatively with the monomer [46]. In order to prepare samples of low polydispersity the polymerization process is carried out preferably in the bulk and around 0–10 °C. The process is stopped when the chains have reached the desired average molecular weight. This is long before depropagation might influence the molecular weight distribution. Deactivation is performed by the addition of an efficient nucleophile such as an alcoholate, a phenoxide [47], an amine, or a phosphine [48], or even lithium bromide. Deactivation by means of sodium phenoxide (Saegusa's method [47]) is most commonly applied.

The general reaction scheme can be written as

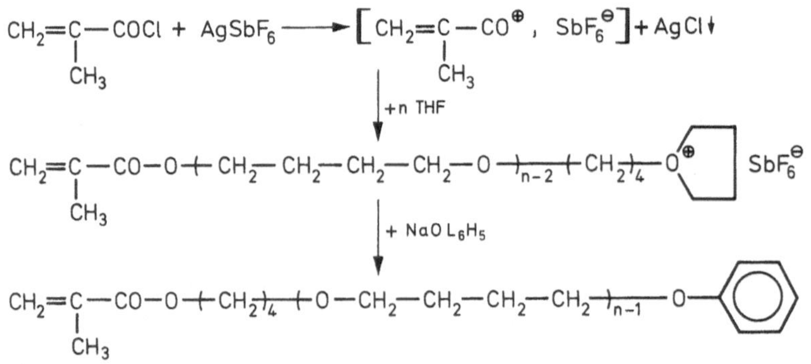

$$CH_2{=}\underset{CH_3}{\overset{|}{C}}{-}COCl + AgSbF_6 \longrightarrow \left[CH_2{=}\underset{CH_3}{\overset{|}{C}}{-}CO^{\oplus}, SbF_6^{\ominus}\right] + AgCl\downarrow$$

$$\downarrow {+}n\ THF$$

$$CH_2{=}\underset{CH_3}{\overset{|}{C}}{-}CO{-}O{+}CH_2{-}CH_2{-}CH_2{-}CH_2{-}O{+}_{n-2}{+}CH_2{+}_4{-}O\overset{\oplus}{\langle\rangle}\ SbF_6^{\ominus}$$

$$\downarrow {+}\ NaO\ C_6H_5$$

$$CH_2{=}\underset{CH_3}{\overset{|}{C}}{-}CO{-}O{+}CH_2{+}_4{+}O{-}CH_2{-}CH_2{-}CH_2{-}CH_2{+}_{n-1}{-}O{-}\langle C_6H_5\rangle$$

A characterization of the obtained samples showed that each molecule contains one ester function, one phenyl ring and one methacrylic unsaturation. The accessible molecular weight range is broad. However, to ensure a high accuracy of the techniques used for the characterization of these macromonomers (NMR, UV, IR spectrometry, vapor pressure osmometry, double bond analysis) and to employ the latter in graft copolymer synthesis molecular weights in the range 1000 to 10000 g · mol^{-1} have been preferred [46].

Another type of cationic initiation procedure starting from an unsaturated compound has been proposed [49]. 3-Phenylallyl bromide is reacted in THF solution with AgSbF$_6$ whereby a poly-THF macromonomer bearing a styryl residue at the chain end is formed:

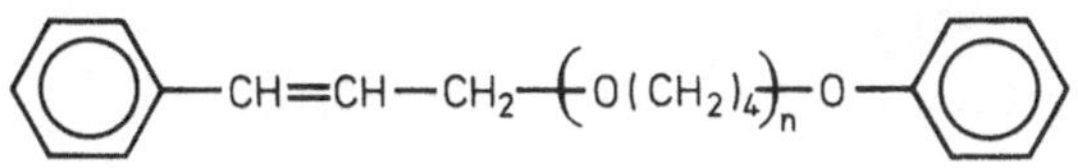

Similar species are formed via cationic deactivation as we shall see in the next section.

2.3.1.2 End-Capping of Living Polymeric Cations

The use of an unsaturated nucleophile to introduce into the chain end of the macromolecule a double bond has also proved successful for the synthesis of poly-THF macromonomers. The *oxolane* polymerization is started with any efficient initiator. When the growing chains have reached the desired length, the unsaturated deactivator is added. The reaction between the oxonium sites and the nucleophile should be fast and free of side reactions. Various unsaturated nucleophiles have been employed, e.g. p-vinylphenoxide used by Asami [50]. The THF polymerization was initiated with triethyloxonium tetrafluoroborate and carried out at 0 °C. Addition of the nucleophile (obtained by reaction of the phenol with NaH) yields the corresponding macromonomer the structure of which was characterized by various techniques:

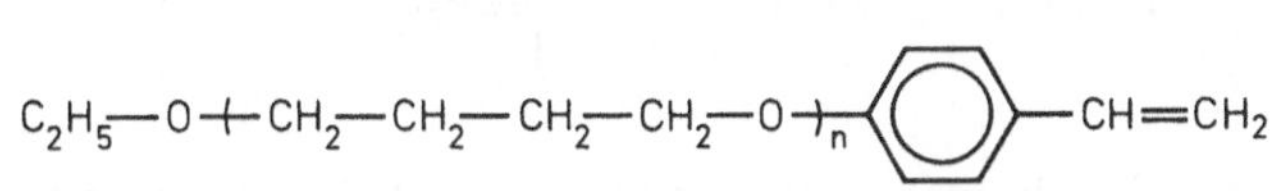

This method was directly derived from Saegusa's deactivation method [47].

— p-Vinyl (or p-isopropenyl) benzylates [51, 52] were also successfully used. The reaction of alkoxides with the cationic living sites is as efficient as that of phenoxide:

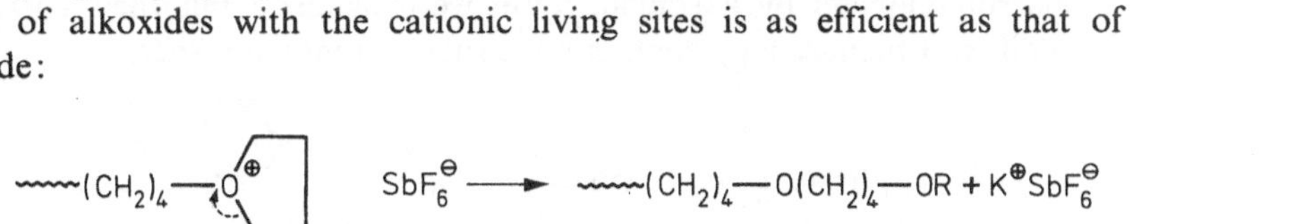

The macromonomers obtained cover a molecular weight range from 2000 to 10000 g/mol and their structure has been determined by means of NMR, UV, GPC, vapor pressure osmometric, and light scattering measurements as well as by double bond analysis:

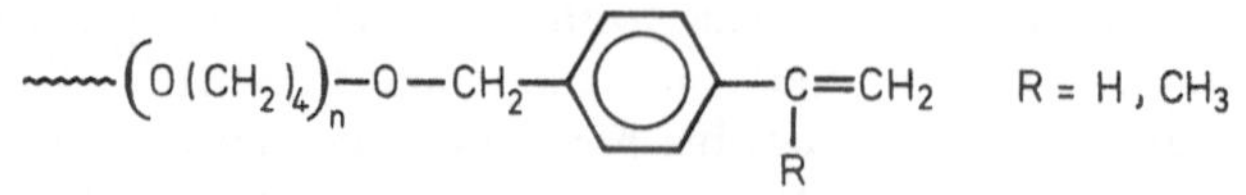

$$\sim\sim\left(O(CH_2)_4\right)_n\!\!-O-CH_2-\!\!\bigcirc\!\!-C=CH_2 \qquad R = H, CH_3$$

$$\underset{R}{\overset{}{|}}$$

It has been detected that each molecule contains a terminal p-vinylbenzyl or p-iso-propenylbenzyl group.

— The lithium salt of cinnamyl alcohol [53] was also shown to be a possible deacti-vator for living poly-THF:

$$\sim\sim O(CH_2)_4\!\!-O^{\oplus}\!\!\bigcirc + Li-O-CH_2-CH=CH-C_6H_5 \xrightarrow{-LiSbF_6}$$

$$\underset{SbF_6^{\ominus}}{}$$

$$\sim\sim O(CH_2)_4-O(CH_2)_4-O-CH_2-CH=CH-C_6H_5$$

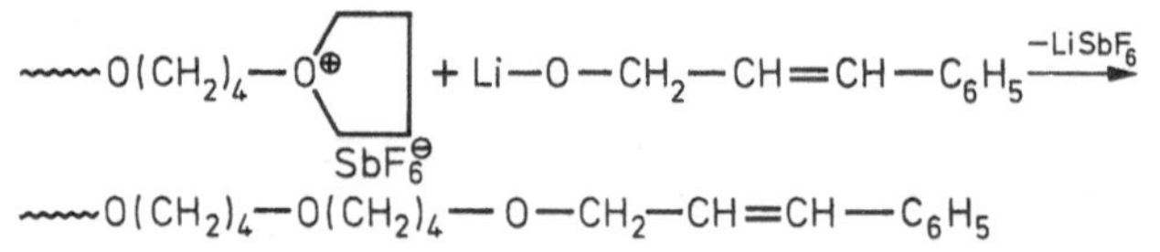

However, in this example the terminal double bond is 1,2-disubstituted and its reac-tivity in free-radical copolymerization can be expected to be rather low. In fact, Richards and Schue [53] studied this process with the purpose of performing a "transformation reaction" [54] from cationic to anionic, the anionic site being generated by the addition of butyllithium to the terminal cinnamyl double bond. However, this addition also occurs in rather low yields.

Even sodium methacrylate [55] can act as an unsaturated deactivator for living poly-THF. The reaction is slow but by a proper choice of the experimental conditions the macromonomers can be obtained in high yields [55] without any change in molecular weight (with respect to the precursor polymer deactivated with sodium phenolate). The macromonomers thus obtained contain a methacrylic ester function at the chain end. They are identical with those prepared by cationic initiation with methacryloyl hexafluoroantimonate:

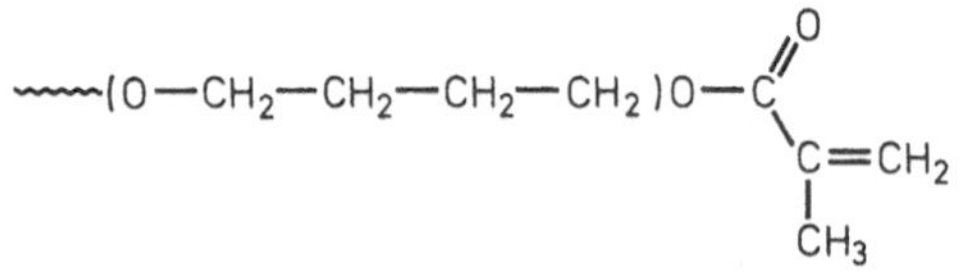

$$\sim\sim(O-CH_2-CH_2-CH_2-CH_2)O-\overset{\overset{O}{\parallel}}{C}$$

$$\underset{CH_3}{\overset{}{\underset{|}{}}}C=CH_2$$

— Amine functions readily react with oxonium sites. This reaction was used for the synthesis of graft copolymers involving reaction of living poly-THF with either poly(p-vinylpyridine) or poly(p-dimethylaminostyrene) [56]. Attempts to synthesize macromonomers by a similar route were made [57] but their characterization is difficult because they contain quaternary ammonium sites:

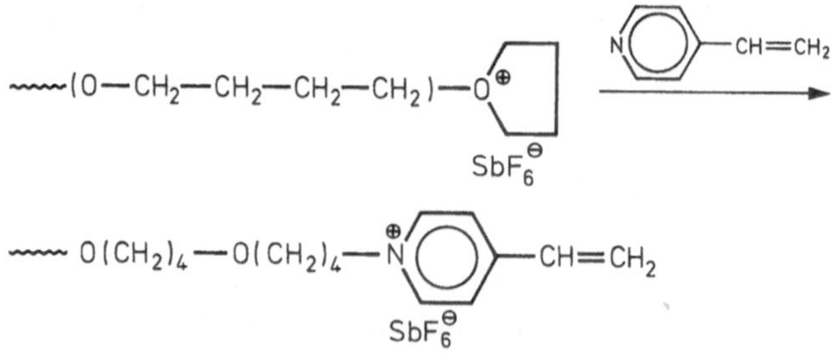

$$\sim\sim(O-CH_2-CH_2-CH_2-CH_2)-O^{\oplus}\!\!\bigcirc \qquad \overset{N}{\bigcirc}\!\!-CH=CH_2$$

$$\underset{SbF_6^{\ominus}}{} \qquad\qquad \longrightarrow$$

$$\sim\sim O(CH_2)_4-O(CH_2)_4-\overset{\oplus}{N}\!\!\bigcirc\!\!-CH=CH_2$$

$$\underset{SbF_6^{\ominus}}{}$$

All these cationic deactivation processes were performed with oxolane as the monomer and with various initiators such as triethyloxonium tetrafluoroborate and benzoyl, acetyl or propionyl hexafluoroantimonate. Efficient difunctional cationic initiators such as adipoyl- or terephthaloyl hexafluoroantimonate) can also be used [42] to synthesize bifunctional macromonomers containing at both chain ends a polymerizable double bond.

The cationic polymerization of *t-butylaziridine* can be characterized as "living": molecular weights are predictable and molecular weight distributions narrow [58]. Goethals showed that the active species readily reacts with nucleophiles; thus, a possibility of synthesizing macromonomers is offered [59]. Methyl trifluoromethane sulfonate was used as the initiator and the polymerization was carried out in a THF/ HMPA mixture. When the polymerization is completed the aziridinium ions are deactivated by the addition of a tenfold excess of methacrylic acid. The corresponding anions are formed by reaction with the amine functions distributed along the chain. After neutralization the polymer is recovered and adequately characterized. The molecular weight calculated from the [monomer]/[initiator] molar ratio satisfactory by agrees with the values determined by vapor pressure osmometry and by NMR whereby the ratio of the methyl groups originating from the initiator to the methacrylic ester functions was found to be close to unity. This indicates that the macromonomer exhibits the expected structure:

$$CH_3-N-CH_2-CH_2\!\!-\!\!\!\left[\!N-CH_2-CH_2\!\!\right]_{n-1}\!\!\!-O-CO-C\!\!\diagdown\!\!\!{\genfrac{}{}{0pt}{}{CH_2}{CH_3}}$$

Again one has the possibility to make bifunctional macromonomers.

Triflic anhydride is an efficient initiator used in the living polymerization of oxolane [60], and bifunctional poly-THF oligomers can be further employed to initiate the polymerization of t-butylaziridine yielding polymers with aziridinium sites at both ends of their chain [61] which can be reacted with methacrylic acid.

2.3.1.3 Cationic Transfer Reactions

Mention should be made of a process first described by Watanabe et al. [62] in which cationic polymerization of alkylene oxides was initiated by Lewis acids and carried out in the presence of methacrylic acid or 2-hydroxyethyl methacrylate (HEMA). The products obtained were characterized and found to contain one terminal methacrylic ester function per chain:

$$CH_2\!\!=\!\!C-CO-O\!\!\left[\!CH_2-CH-O\!\right]_n\!\!H \qquad R = H, CH_3$$
$$CH_3 R$$

---

1 Attempts to synthesize the same type of compounds by anionic polymerization of oxiranes in the presence of transfer agents such as HEMA failed because anionic polymerization requires temperatures at which thermally initiated polymerization of the methacrylic double bond occurs

The advantage of this procedure is that it yields monofunctional poly(alkylene oxide) macromonomers whereas a drawback is the low degree of polymerization (p < 20)[1].

Another method involving cationic transfer was investigated by Heitz et al. [63,64] who performed the ring-opening polymerization of oxolane in the presence of acrylic or methacrylic anhydride with either triflic acid (CF$_3$SO$_3$H) or acetyl hexafluoroantimonate as the initiator. Once the monomer has reached its equilibrium concentration a kind of reshuffling involving the anhydride takes place whereby the molecular weight of the poly-THF sharply decreases, and methacrylic ester functions are introduced into the polymer molecules. It was found that the reaction is not in accordance with the postulated mechanism:

$$\sim\!\!\sim\!\!(CH_2)_4\!-\!\overset{\oplus}{O} \quad + \quad O\!\!\begin{array}{c} \nearrow COR \\ \searrow COR \end{array} \xrightarrow{\quad\times\quad} \sim\!\!\sim\!\!(CH_2)_4\!-\!O(CH_2)_4\!-\!\overset{\oplus}{O}\!\!\begin{array}{c} \nearrow COR \\ \searrow COR \end{array}$$

$$\xrightarrow{\quad\times\quad} \quad R\overset{\oplus}{C}O + \sim\!\!\sim\!\!(CH_2)_4\!-\!O(CH_2)_4\!-\!O\!-\!CO\!-\!R$$

It seems that instead protonated species (anhydride or ester molecules) play a major role in the process. The protons originate from some added acid (e.g. acrylic or methacrylic acid). The characterization of the formed macromonomers revealed that the number of ester functions per molecule is close to 2. The role of the protons is evidenced by the increase of the reaction rate with increasing amount of methacrylic acid in the system. In the absence of a protonic acid high molecular weight poly-THF is produced, no anhydride is consumed and reshuffling does not take place. This mechanism which remains to be confirmed is in any case completely different from the "infer"-type cationic transfer which may occur with unsaturated monomers. It is discussed in the next section.

The preparation of PDMS by Katz (see "formation of vinylsilane") should be mentioned here.

2.3.2 Vinylic Monomers: "Inifer"-Type Methods of Synthesizing Macromonomers

The cationic polymerization of vinyl monomers such as isobutene, styrene, α-methylstyrene, indene, and vinyl ethers is generally impeded by several transfer processes [65]. The molecular weight of the polymer is not determined by the molar ratio of monomer to initiator. Therefore, the methods that are based on the long lifetime of the active sites at the chain ends cannot be applied here.

Nevertheless, a few years ago, Kennedy [66-69] developed a method yielding ω-functional polymers by cationic polymerization of vinyl monomers. The principle of the socalled "inifer" method is to kinetically favor transfer to the initiating species with respect to all other kinds of transfer reactions (especially the transfer to monomer). A typical initiating system is composed of an allyl or benzyl halide and boron trichloride BCl$_3$. This mixture behaves like an alkenium tetrachloroborate and readily initiates the polymerization of monomers such as isobutene or α-methylstyrene. The efficiency of the halide as a transfer agent depends on the lability of the C—Cl bond and on the molar ratio [RCl]/[BCl$_3$].

The proposed reaction scheme is as follows:

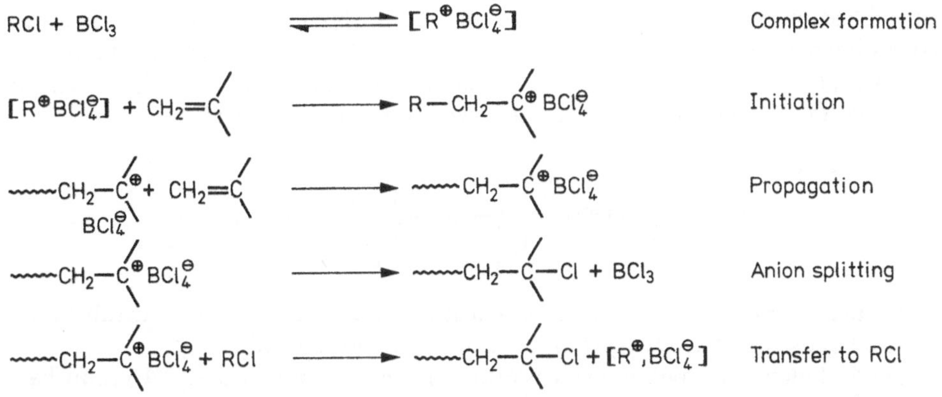

$$RCl + BCl_3 \rightleftharpoons [R^{\oplus}BCl_4^{\ominus}] \qquad \text{Complex formation}$$

$$[R^{\oplus}BCl_4^{\ominus}] + CH_2=C\diagdown \longrightarrow R-CH_2-C^{\oplus}\diagup BCl_4^{\ominus} \qquad \text{Initiation}$$

$$\sim\!\!\!\sim CH_2-C^{\oplus}\diagup + CH_2=C\diagdown \longrightarrow \sim\!\!\!\sim CH_2-C^{\oplus}\diagup BCl_4^{\ominus} \qquad \text{Propagation}$$
$$BCl_4^{\ominus}$$

$$\sim\!\!\!\sim CH_2-C^{\oplus}\diagup BCl_4^{\ominus} \longrightarrow \sim\!\!\!\sim CH_2-C\diagup-Cl + BCl_3 \qquad \text{Anion splitting}$$

$$\sim\!\!\!\sim CH_2-C^{\oplus}\diagup BCl_4^{\ominus} + RCl \longrightarrow \sim\!\!\!\sim CH_2-C\diagup-Cl + [R^{\oplus},BCl_4^{\ominus}] \qquad \text{Transfer to RCl}$$

Since the same species (RCl) acts as initiator *and* transfer agent this type of process is referred to as "inifer".

In such systems transfer to monomer is negligible. The average molecular weight theoretically results from the ratio of the propagation rate to the rate of transfer; experimentally, it is governed by the monomer-to-inifer mole ratio. An interesting feature of this type of process is that anion splitting, if it occurs, yields exactly the same species as the regulár transfer to the organic halide, as it can be seen from the above scheme.

"Inifer"type reactions have been applied to a large number of cationic synthesis, yielding functional polymers [70-72], block copolymers [73], graft copolymers, and even star-shaped polymers [73]. As far as macromonomer synthesis is concerned three different methods have been used which are described below:

a) Polyisobutene bearing a tertiary chlorine atom at the chain end can be readily obtained by using, for instance, the efficient system cumyl chloride boron trichloride. Subsequent dehydrochlorination creates a terminal double bond [74]. However, both internal and terminal double bonds are obtained [75]. The macromonomers have been characterized but their ability to copolymerize has been given little attention:

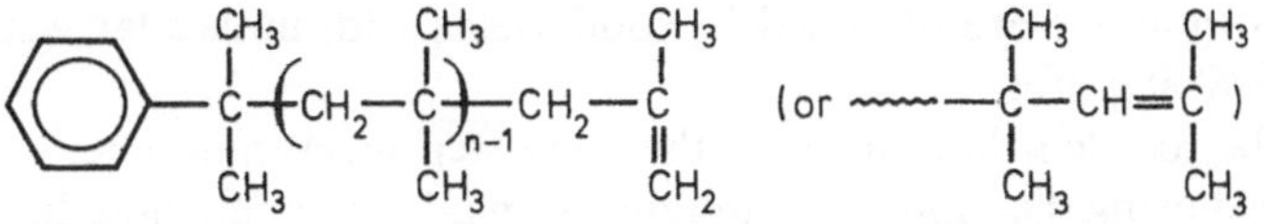

When dicumenyl chloride is used as the inifer it was established [75] that part of the molecules formed was monofunctional instead of bifunctional. This is due to an intramolecular cyclization reaction yielding an indanyl group after the addition of only one isobutene unit:

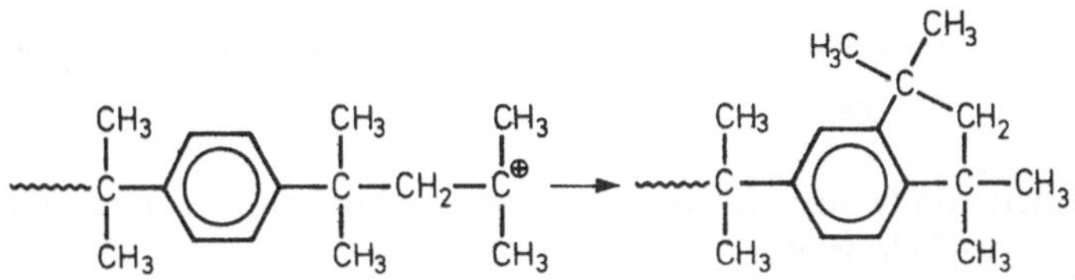

It can be expected that this side reaction can also occur with cumyl chloride, whereby the macromonomer yield should be lowered.

b) ω-Unsaturated polyisobutenes have also been obtained by means of a system consisting of a substituded allyl chloride and BCl_3 [76]. Characterization data indicate that the species contain one double bond per molecule, the molecular weights being of the order of 10000 g/mol. However, this type of macromonomer

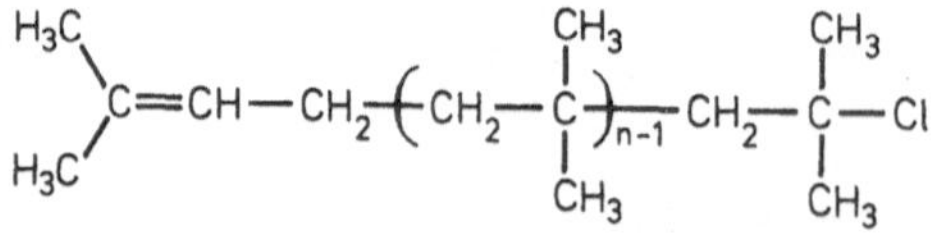

is practically unimportant as far as applications are concerned, since trisubstituted double bonds are relatively inert in free-radical copolymerization.

c) Polyisobutene macromonomers bearing at their chain end a styryl group have also been synthesized by Kennedy [69, 77]. The initiating species consists of p-vinyl-benzyl chloride [69], triethylaluminium and water. The reaction was carried out in methylene chloride at low temperature. Assuming the reaction scheme described above, the macromonomer formed should exhibit the following structure:

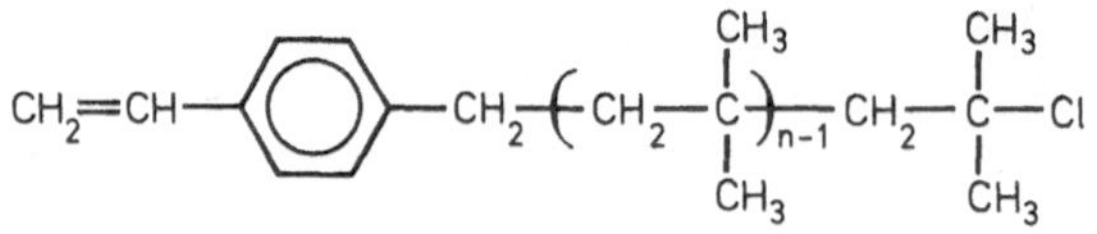

Whether all chains bear a terminal vinylic double bond has not been clearly established, and it would be somewhat astonishing if vinylic double bonds did not undergo side reactions since their reactivity in cationic polymerization is quite high. However, the occurrence of terminal p-vinyl benzyl groups is confirmed by the fact that the formed macromonomer readily copolymerizes with butyl acrylate.

An improved method [77], though involving two steps, makes use of p-(β-bromo-ethyl)cumenyl chloride as the "inifer", thus leading to functional precursor poly-isobutene with a β-bromoethylphenyl group at the chain end. Dehydrobromination of this group is performed in a later step whereby macromonomers exhibiting the structure shown above are obtained in good yields with molecular weights ranging from 1000 to 4000 g/mol.

It should be recalled that various other two-step macromonomer synthesis have been developed using the "inifer" technique as an adequate functionalization method for preparing the precursor. Thus, starting from ω-chloro-polyisobutene, the de-hydrohalogenation can be followed by hydroboration and selective oxidation to yield almost quantitatively the ω-hydroxy precursor [78]:

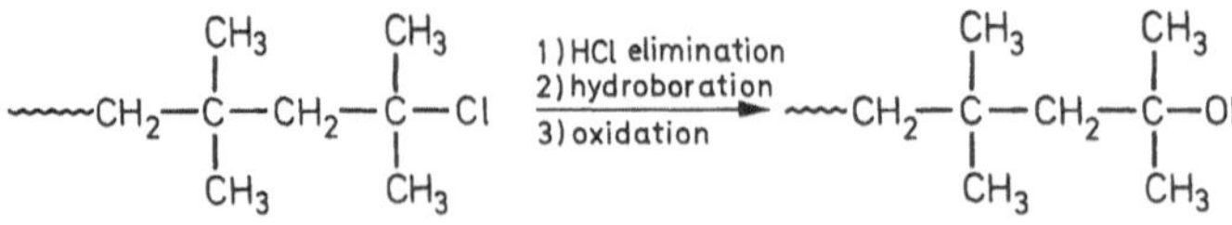

These polymers can be used for the preparation of macromonomers, e.g. by reacting them with acryloyl chloride.

Other functionalizations of polyisobutene have been performed [78] by means of the "inifer" technique and subsequent reaction at the chain end. Polymers bearing terminal carboxylic acid or anhydride functions can also be considered as potential starting materials for the synthesis of polyisobutene macromonomers.

2.4 Macromonomer Synthesis by Addition Processes

In 1973, Tsuruta and his coworkers [79] established that diethylamine adds stoichiometrically to styrene in the presence of some lithium diethylamide. The reaction mechanism involves two steps namely a kind of anionic initiation and subsequent proton transfer, due to the difference in the acidity between the benzylic and amine substrate:

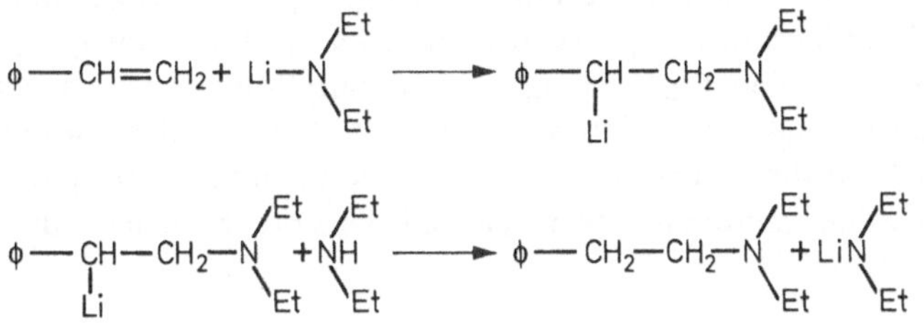

$$\phi\text{---CH=CH}_2 + \text{Li---N}\begin{smallmatrix}\text{Et}\\\text{Et}\end{smallmatrix} \longrightarrow \phi\text{---CH---CH}_2\text{---N}\begin{smallmatrix}\text{Et}\\\text{Et}\end{smallmatrix}$$

$$\phi\text{---CH---CH}_2\text{---N}\begin{smallmatrix}\text{Et}\\\text{Et}\end{smallmatrix} + \text{NH}\begin{smallmatrix}\text{Et}\\\text{Et}\end{smallmatrix} \longrightarrow \phi\text{---CH}_2\text{---CH}_2\text{---N}\begin{smallmatrix}\text{Et}\\\text{Et}\end{smallmatrix} + \text{LiN}\begin{smallmatrix}\text{Et}\\\text{Et}\end{smallmatrix}$$

Also 1,3-dienes undergo such an addition reaction; the adducts formed almost exclusively exhibit cis-1,4 structure [80].

When applied to p-divinylbenzene (DVB), this reaction exhibits an interesting feature [81]: a monoadduct is easily obtained and isolated because of the rather large difference in the reactivity between the two double bonds of p-DVB. However, when the reaction time is increased, the second DVB double bond may also be involved in the addition. The rate of the first step is roughly 20 times faster than that of the second:

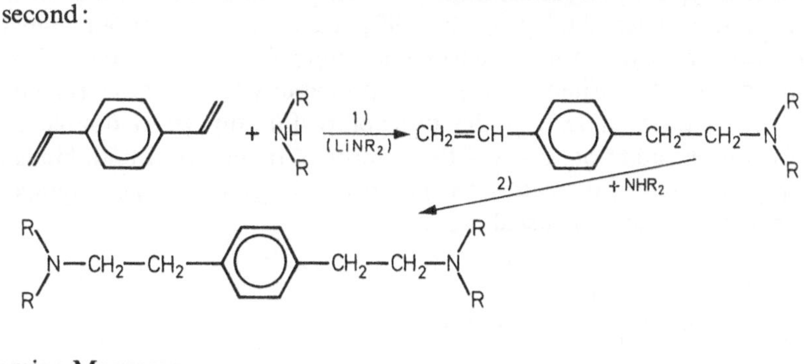

2.4.1 Polyamine Macromonomers

This reaction was applied to the synthesis of macromonomers [81-83]. If p-DVB is reacted with N,N'-diethylethylenediamine in the presence of a lithiated diamine the reaction is fast and a monoadduct is formed provided stoichiometric proportions of the two reactants are used. The monoadduct can be isolated and purified:

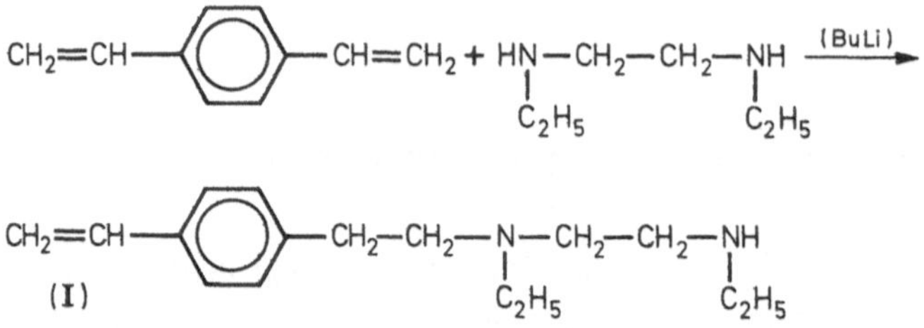

If the diadduct is desired, a large excess of the diamine is required, and the reaction mixture has to be kept at $-17\,°C$ for a long period to attain high conversion.

The monoadduct (I) contains a terminal vinyl group and a secondary amine function. Thus, self-condensation appears to be possible. Lithium diisopropylamide was used to create the amide function, instead of butyllithium which could possibly attack the double bond. Thus, the metalation of the secondary amine function is straightforward, and the occurring self-condensation can be monitored by ^{13}C-NMR and by GPC. No side reaction has been detected. This process can be illustrated by the following scheme:

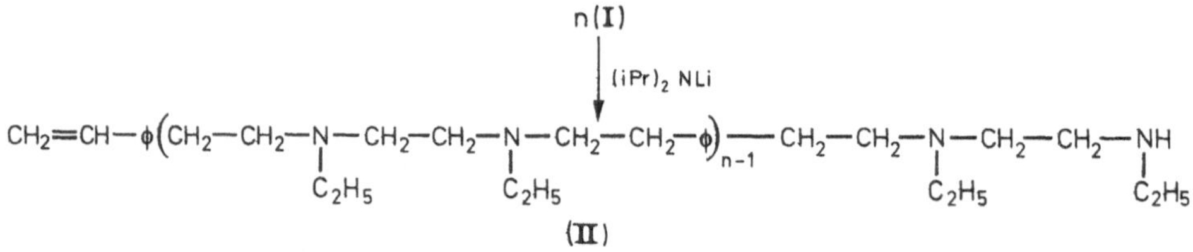

It is a non-typical polycondensation involving reaction of an i-mer with a j-mer to yield an (i + j)mer similar to the self-polycondensation of an ω-hydroxy acid. Kinetically, the reaction also behaves like a step polymerization and yields a Gaussian molecular weight distribution. If p is the probability for any given (metalated) amine function to react, it is also the probability for any given double bond to get involved in the addition, because the number of functions of both kinds is the same at any time. Hence, the calculated number average and weight average degrees of polymerization using the usual expression

$$X_n = \frac{1}{1-p} \qquad X_w = \frac{1+p}{1-p}$$

are in excellent agreement with the experimental values.

However, each step of the reaction can also be considered as an anionic polymerization that does not propagate. The stabilized lithiated site (III), resulting from the addition to a double bond, is unable to attack another double bond, and can only undergo protonation by the more acidic N,N'-diethylethylenediamine.

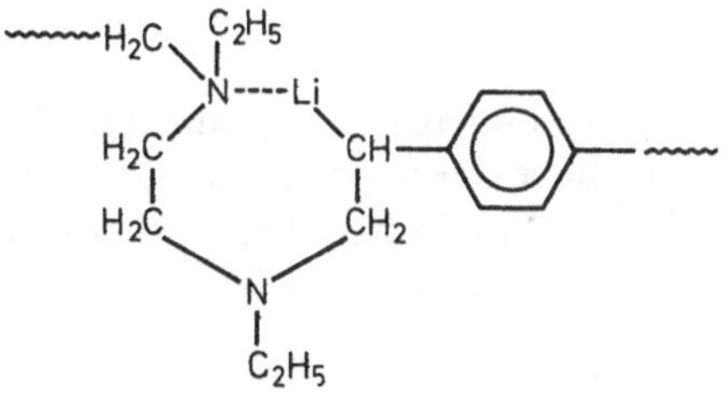

III

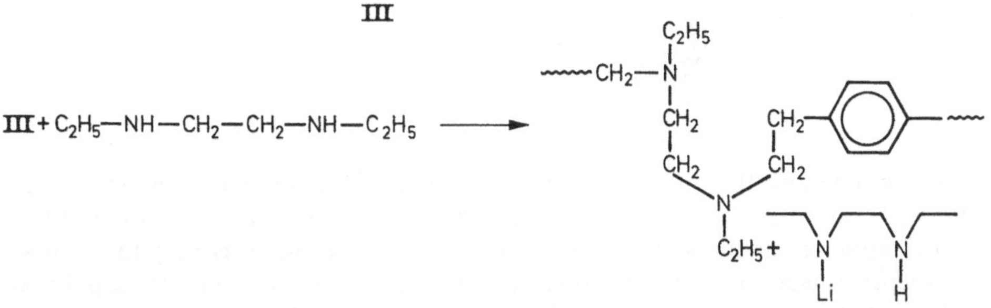

$$III + C_2H_5-NH-CH_2-CH_2-NH-C_2H_5 \longrightarrow$$

An adequate control of the average molecular weight of the macromonomers formed can be achieved. Macromonomers of the type (II) covering a wide range of molecular weights have been obtained and characterized.

The method of Tsuruta was also extended to the synthesis of macromonomers containing piperazine cycles, and even diaza crown ethers [81, 84, 85]. Here again the first step involves preparation of 1:1 adducts of p-divinylbenzene and the corresponding heterocyclic compound:

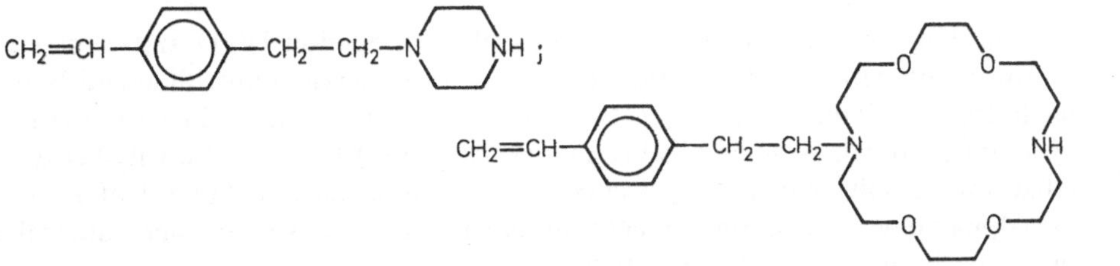

Self-condensation of these adducts yields macromonomers exhibiting the following structures:

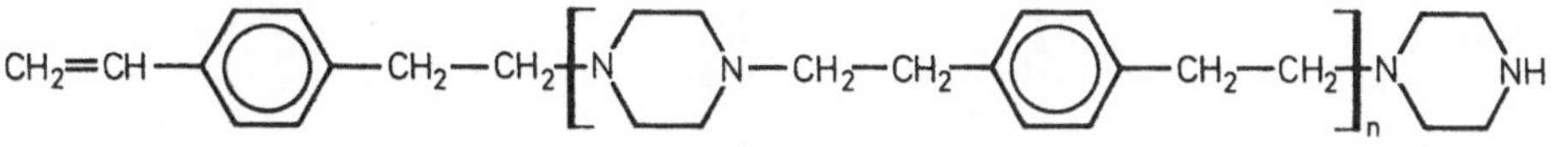

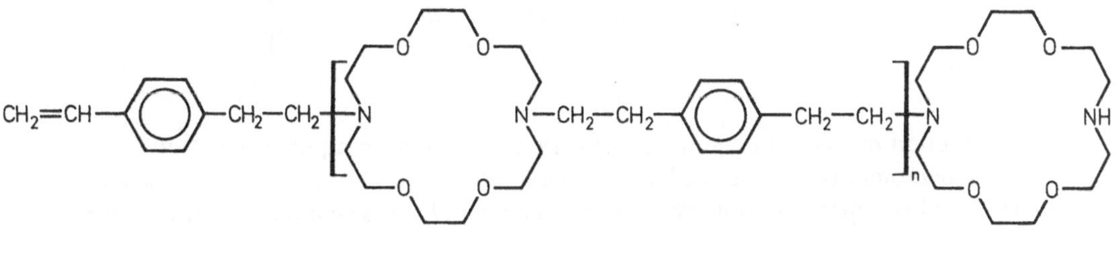

These two macromonomers were subsequently copolymerized with styrene to yield graft copolymers [84] containing crown ethers in their side chains.

2.4.2 Polypeptide Macromonomers

The monoadduct of p-DVB and N,N'-diethylethylenediamine (I) is itself a new monomer. Several other monomers have been prepared by reaction of p-DVB with various amines or diamines [82, 86]. p-Vinylbenzyl derivatives carrying a primary amine function (instead of a secondary amine moiety) have also been obtained in high yields [87]:

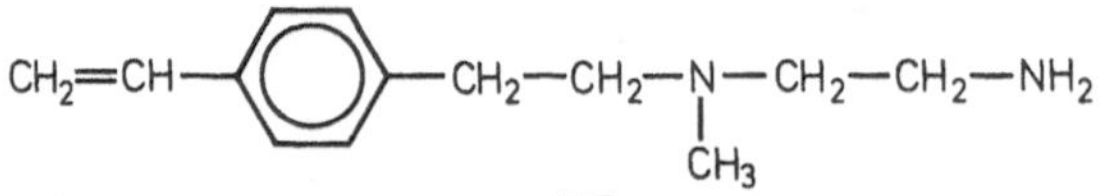

$$CH_2{=}CH{-}\bigcirc{-}CH_2{-}CH_2{-}\underset{\underset{CH_3}{|}}{N}{-}CH_2{-}CH_2{-}NH_2$$

$$\text{IV}$$

It is well known that primary amines are efficient initiators for the polymerization of Leuch's anhydrides (oxazolidinediones) and that initiation proceeds by the addition of the amine to the monomer. This pathway has been utilized recently to synthesize polypeptide macromonomers bearing a terminal p-vinylbenzyl group [88]. Copolymerization of these macromonomers with a vinylic or acrylic comonomer yields graft copolymers with polypeptide grafts. Alternately, the monomer adduct (IV) was copolymerized with styrene, and the primary amine functions of this polymer were used to initiate the polymerization of an oxazolidinedione whereby polypeptide grafts are formed [89]. Such graft copolymers may be of interest for biomedical applications.

2.4.3 Bis-crylamide Macromonomers

The same kind of addition reaction as described in Section IV 1) and 2) can be performed with substituted bisacryl amides such as bisacryloylpiperidide and N,N'-dimethylethylenediamine. However, the difference in the reactivity between the two acrylic double bonds is much less pronounced than in the case of p-divinylbenzene, so that a mere polycondensation occurs. Upon using a calculated excess of 1,4-bis-acryloylpiperidine bifunctional macromonomers of a known average molecular weight have been obtained by Ferruti et al. [90]:

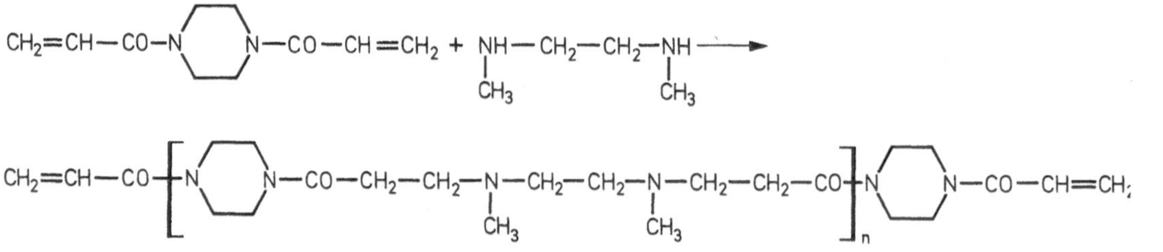

$$CH_2{=}CH{-}CO{-}N\bigcirc N{-}CO{-}CH{=}CH_2 + \underset{\underset{CH_3}{|}}{NH}{-}CH_2{-}CH_2{-}\underset{\underset{CH_3}{|}}{NH} \longrightarrow$$

$$CH_2{=}CH{-}CO{-}\Big[N\bigcirc N{-}CO{-}CH_2{-}CH_2{-}\underset{\underset{CH_3}{|}}{N}{-}CH_2{-}CH_2{-}\underset{\underset{CH_3}{|}}{N}{-}CH_2{-}CH_2{-}CO\Big]_n N\bigcirc N{-}CO{-}CH{=}CH_2$$

This reaction proceeds at room temperature in two days in aqueous solution.

The macromonomer obtained was copolymerized with styrene in the presence of AIBN and is apparently not cross-linked. The copolymers obtained seem to exhibit good bio-compatibility and may be used, after reaction with heparine, as long acting non-thrombogenic biomaterials.

2.5 Other Two-Step Methods for the Synthesis of Macromonomers

The reaction of an unsaturated compound with an antagonist function located at the end of a polymer chain is still the most commonly used method to synthesize macromonomers. We have already mentioned some processes that can be used to introduce into the chain end of a macromolecule a functional group, e.g. by deactivation of living carbanionic sites and transfer reactions of various kinds in cationic polymerization. We have also described some methods used to link an active terminal double bond to the chain end originally bearing hydroxy groups.

We shall discuss in this section alternatives for synthesizing quantitatively ω-functional polymers and which allow at least some kind of control of the molecular weight of the species formed. These polymers can thus serve as precursors for macromonomer synthesis.

2.5.1 Methods Involving Step Polymerization

Polycondensation reactions involving two types of bifunctional components AA and BB (A and B are antagonist functions) are unimportant for our purpose. If the components are reacted stoichiometrically to high conversion, the polycondensate molecules carry A and B functions at their chain end in equal amounts which are however randomly distributed: 25% A———A, 25% B———B, 50% A———B molecules.

To obtain polymer chains bearing at both ends the same function, say A, it is necessary to carry out to high conversion the reaction of a non-stoichiometric mixture of AA and BB. This procedure also allows some control of the molecular weight of the polycondensate since the well-known equation

$$\overline{DP}_n = \frac{1 - r}{1 + 2p(1 - r) - r}$$

applies.

where $r = \dfrac{[BB]}{[AA]} \leqq 1$ and p = degree of conversion.

Such α-ω-bifunctional polycondensates are only of interest for the synthesis of bifunctional macromonomers, and, to our knowledge, no research has been carried out along this line.

The self-polycondensation of molecules bearing two antagonist functions at their chain ends (denoted as AB) is more interesting since at any stage of the reaction, the polymer molecules always bear one A and one B function at the chain end. If one of these functions can be reacted subsequently with an unsaturated compound, the synthesis of macromonomers becomes possible.

This procedure was chosen by Waite [91] to synthesize polyester macromonomers starting from 12-hydroxystearic acid. The self-condensation of this compound was carried out to the desired conversion p which also determines the average degree of polymerization; thereafter, the terminal carboxylic groups were reacted with glycidyl methacrylate, thus yielding the macromonomer:

$$H\text{---}\left(O\text{---}\underset{\underset{C_6H_{13}}{|}}{CH}\text{---}(CH_2)_{10}\text{---}CO\right)_{n-1}\text{---}O\text{---}\underset{\underset{C_6H_{13}}{|}}{CH}\text{---}(CH_2)_{10}\text{---}COOH + CH_2\text{---}\underset{\underset{O}{\diagdown\diagup}}{C}\text{---}CH_2\text{---}O\text{---}CO\text{---}\underset{\underset{CH_3}{|}}{C}\text{=}CH_2$$

$$H\text{---}\left(O\text{---}\underset{\underset{C_6H_{13}}{|}}{CH}\text{---}(CH_2)_{10}\text{---}CO\right)_{n-1}\text{---}O\text{---}\underset{\underset{C_6H_{13}}{|}}{CH}\text{---}(CH_2)_{10}\text{---}CO\text{---}O\text{---}CH_2\text{---}\underset{\underset{OH}{|}}{CH}\text{---}CH_2\text{---}O\text{---}CO\text{---}\underset{\underset{CH_3}{|}}{C}\text{=}CH_2$$

Polyester macromonomers were also prepared by Hudecek et al. [92] who reacted polyesters with hydroxy end groups. With the mono-adduct of 2-hydroxyethyl methacrylate (HEMA) and methylene-bis-phenylene isocyanate (MDI) one obtains:

$$\text{\~\~\~Polyester\~\~\~OH} + O\text{=}C\text{=}N\text{---}\phi\text{---}CH_2\text{---}\phi\text{---}NH\text{---}CO\text{---}O\text{---}CH_2\text{---}CH_2\text{---}O\text{---}CO\text{---}\underset{\underset{CH_3}{|}}{C}\text{=}CH_2$$

$$\text{\~\~\~Polyester\~\~\~O}\text{---}CO\text{---}NH\text{---}\phi\text{---}CH_2\text{---}\phi\text{---}NH\text{---}CO\text{---}O\text{---}CH_2\text{---}CH_2\text{---}O\text{---}CO\text{---}\underset{\underset{CH_3}{|}}{C}\text{=}CH_2$$

At this stage, the self-condensation of the p-divinylbenzene-N,N'-ethylenediamine monoadduct [83, 84] should be mentioned again as the only type of self-condensation leading directly to a macromonomer without any additional reaction: the final species always carries an unsaturation at one chain end.

2.5.2 Methods Involving Free-Radical Polymerization

The main feature of free-radical polymerization is the very short lifetime of the growing radicals. Consequently, functionalization can arise only from the use of functional initiators or from transfer processes [93].

2.5.2.1 Functional Initiators

A radical formed upon homolytic cleavage of a functional initiator carries the corresponding function. Since upon initiation the primary radical adds to a monomer, this functional group remains attached to the formed polymer molecule. However, this functionalization procedure is far from being adequate for the following reasons:
— Functionalization at only one chain end requires that termination occurs exclusively by disproportionation (and not by recombination) and that transfer reaction of the radical to monomer and to solvent molecules can be disregarded. If the first condition is not fulfilled, functionalization at both chain ends can result. If the second condition is not valid, only a fraction of the formed macromolecules is functionalized, namely those arising from primary radicals.

— A control of the molecular weights of the polymers formed under standard experimental conditions is not possible. It is well known that low molecular weight species can be obtained only in free-radical polymerization involving transfer reactions. Some initiators can act as transfer agents as well. By carrying out polymerization in the presence of a large amount of a functional initiator it is possible to lower the molecular weight of the formed polymer and to control it sufficiently. However, this implies that *both* chain ends of the macromolecules bear functional initiator fragments [94].

Typical examples of functional initiators [95, 96] include azo-bis-4-cyanovaleric acid and azo-bis-4-cyanopentanol

$$HOOC-(CH_2)_2-\underset{\underset{CN}{|}}{\overset{\overset{CH_3}{|}}{C}}-N\!=\!N-\underset{\underset{CN}{|}}{\overset{\overset{CH_3}{|}}{C}}-(CH_2)_2-COOH$$

$$HO(CH_2)_3-\underset{\underset{CN}{|}}{\overset{\overset{CH_3}{|}}{C}}-N\!=\!N-\underset{\underset{CN}{|}}{\overset{\overset{CH_3}{|}}{C}}-(CH_2)_3\,OH$$

the behaviour of which is similar to that of AIBN.

The work of Brosse and Legeay [97] on the introduction of hydroxy groups into both chain ends of polybutadienes by initiation of the polymerization with H_2O_2 should also be mentioned.

It can be asserted that the synthesis of well-defined functional polymers by means of functional free-radical initiators is far from being satisfactory. To our knowledge, this pathway was never used in this simple form as a step in the macromonomer synthesis.

2.5.2.2 Functionalization by Transfer Reactions

The main role of a transfer agent in a free-radical copolymerization is to control the molecular weight of the species formed by a proper choice of the molar ratio [monomer]/[transfer agent]. The efficiency of a transfer agent S for a given monomer is measured by its transfer constant C_s.[2] If the transfer agent is very effective and present in a sufficiently large amount the limiting degree of polymerization is given by the ratio of the propagation rate to the rate of transfer. The molecular weight distribution of the produced polymer at any given time is Gaussian which implies that the M_w/M_n ratio is equal to 2. If care is taken to avoid large changes in the molar ratio of monomer to transfer agent as conversion proceeds, broadening of the molecular weight distribution can be prevented.

Functionalization at *one* chain end can be achieved by means of efficient functional transfer agents. The problem is to find species which simultaneously allow functionalization and adequate control of the molecular weight. A large value of C_s is essential

---

2 The transfer constant C_s of a given substance S is defined as the ratio of the rate constant of transfer to S to the rate constant of propagation. It always refers to a given monomer and to a given temperature.

because it minimizes the amount of polymer produced by primary radicals (originating from the initiator). The higher the probability of the occurrence of a transfer reaction, the lower the proportion of the polymer molecules devoid of a terminal function.

Successful results have been obtained in the following cases:

— The use of 2-mercaptoethanol, $HS-CH_2-CH_2-OH$, as a transfer agent belongs to the early attempts to synthesize macromonomers by means of free-radical polymerization [10]. Methyl methacrylate was employed as the monomer. The formed polymers bear a terminal hydroxy group which was subsequently reacted with methacryloyl chloride.

— Thioglycolic acid, $HS-CH_2-COOH$, is even more efficient as a transfer agent [2, 98, 99]. Polymerization of various monomers such as styrene, methyl methacrylate and even stearyl methacrylate was carried out in the presence of thioglycolic acid whereby a carboxy group is introduced into the chain end of the polymer molecule.

$$\sim\sim CH_2-\overset{\cdot}{CH} + HS-CH_2-COOH \longrightarrow \sim\sim CH_2-CH_2 + \overset{\cdot}{S}-CH_2-COOH$$
$$\qquad\qquad R \qquad\qquad\qquad\qquad\qquad\qquad\qquad R$$

$$HOOC-CH_2-\overset{\cdot}{S} + CH_2{=}CH \longrightarrow HOOC-CH_2-S-CH_2-\overset{\cdot}{CH}$$
$$\qquad\qquad\qquad\qquad\quad R \qquad\qquad\qquad\qquad\qquad\qquad\qquad R$$

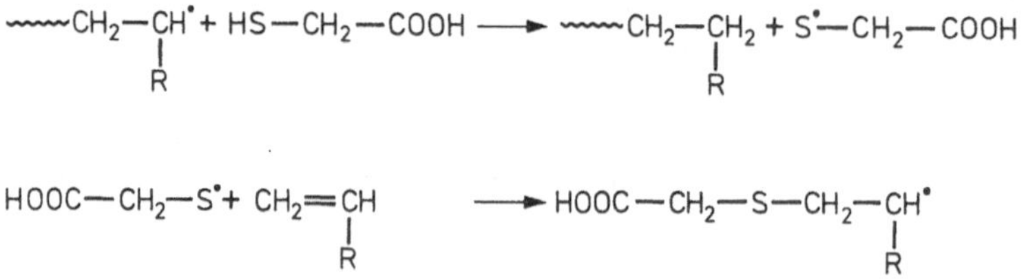

The subsequent step generally involves reaction with glycidyl methacrylate. The resulting macromolecule bears a methacryloyl group at the chain end the olefine double bond of which may be polymerized:

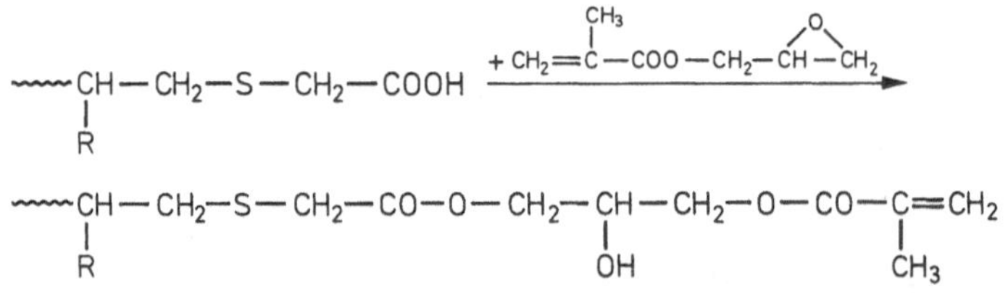

$$\sim\sim CH-CH_2-S-CH_2-COOH \xrightarrow{+ CH_2{=}\overset{CH_3}{\underset{|}{C}}-COO-CH_2-CH-CH_2}$$
$$\qquad R$$

$$\sim\sim CH-CH_2-S-CH_2-CO-O-CH_2-CH-CH_2-O-CO-C{=}CH_2$$
$$\qquad R \qquad\qquad\qquad\qquad\qquad\qquad OH \qquad\qquad\qquad CH_3$$

With stearyl methacrylate [98], using AIBN as the initiator, a satisfactory control of the molecular weight can be achieved by appropriate choice of the molar ratio [monomer]/[thioglycol acid]. It was also established that the subsequent reaction with glycidyl methacrylate does not induce detectable changes in the molecular weight.

A further improvement can be achieved by using a functional initiator [100] such as azo-bis-cyanovaleric acid. The primary radicals, originating from this initiator, and the radicals formed by transfer with thioglycolic acid both carry carboxy groups so that the fraction of the functionalized macromolecules is high.

— Iodoacetic acid [101] was also used as a transfer agent for the polymerization of styrene initiated by AIBN. The efficiency of this transfer agent is relatively high. Thus, 6 mol-% with respect to monomer are sufficient to obtain molecular weights as low as 4,000. In this case, conversions need not to be kept low, and the molecular weights of the prepolymer determined by vapor pressure osmometry and acidimetric titration agree satisfactorily. This implies that the fraction of unfunctionalized macromolecules is very small. Subsequently, the ω-carboxy polymer is reacted with glycidyl methacrylate to yield a polystyrene macromonomer bearing at the chain end a methacryloyl group. The latter reaction was shown to induce no change in the molecular weight of the polymer:

$$\text{I}-\text{CH}-\text{CH}_2(\text{CH}-\text{CH}_2)_n-\text{CH}-\text{CH}_2-\text{CH}_2-\text{COO}-\text{CH}_2-\underset{\overset{|}{\text{OH}}}{\text{CH}}-\text{CH}_2-\text{O}-\text{CO}-\underset{\overset{|}{\text{CH}_3}}{\text{C}}=\text{CH}_2$$

The reaction of terminal acid functions with glycidyl methacrylate is generally considered to proceed readily and quantitatively. However, Yamashita [102] found that some side reactions can occur at elevated temperature, and he proposed another method involving reaction of the terminal carboxy function of the macromolecule with either 2-hydroxyethyl methacrylate or N-methyl-N-2-hydroxyethyl methacrylamide in the presence of bis-isopropylcarbodiimide. This two-step process was shown to proceed quantitatively. The reaction scheme is as follows:

$$\text{CH}_2=\underset{\overset{|}{\text{CH}_3}}{\text{C}}-\text{CO}-\underset{\overset{|}{\text{CH}_3}}{\text{N}}-\text{CH}_2-\text{CH}_2-\text{OH} + \text{HOOC}-\text{CH}_2-\underset{\overset{|}{\text{R}}}{\text{CH}}\sim\sim \xrightarrow{+ \ i\text{Pr}-\text{N}=\text{C}=\text{N}-i\text{Pr}}$$

$$\text{CH}_2=\underset{\overset{|}{\text{CH}_3}}{\text{C}}-\text{CO}-\underset{\overset{|}{\text{CH}_3}}{\text{N}}-\text{CH}_2-\text{CH}_2-\text{O}-\text{CO}-\text{CH}_2-\underset{\overset{|}{\text{R}}}{\text{CH}}\sim\sim$$

Alternately, terminal carboxy groups can also be reacted with aziridines carrying an unsaturated group on the nitrogen atom [103] to synthesize macromonomers in good yields:

$$\underset{\text{H}_2\text{C}}{\overset{\text{H}_2\text{C}}{\diagdown\diagup}}\text{N}-(\text{CH}_2)_x-\text{OCO}-\underset{\overset{|}{\text{CH}_3}}{\text{C}}=\text{CH}_2 + \text{HOOC}-\text{CH}_2-\underset{\overset{|}{\text{R}}}{\text{CH}}\sim\sim \longrightarrow$$

$$\text{CH}_2=\underset{\overset{|}{\text{CH}_3}}{\text{C}}-\text{CO}-\text{O}(\text{CH}_2)_x-\text{NH}-\text{CH}_2-\text{CH}_2-\text{O}-\text{CO}-\text{CH}_2-\underset{\overset{|}{\text{R}}}{\text{CH}}\sim\sim$$

$$x = 1 \ldots 8$$

The macromonomers thus obtained were subsequently copolymerized with various comonomers.

It should be mentioned [104] that the aziridinyl methacrylate can be copolymerized first with a comonomer, and subsequently the dangling aziridinyl moieties are reacted with short polymer chains bearing terminal carboxy groups. Both procedures yield the same graft copolymers.

Macromonomers used for stepwise polymerizations, i.e. polymer chains bearing two hydroxy or two carboxy groups at one chain end can also be obtained by free-radical transfer reactions using bifunctional transfer agents.

This is exemplified by the polymerization of methyl methacrylate carried out in the presence of thiomalic acid [105] which is an efficient transfer agent. The reaction scheme can be written as:

$$
\begin{array}{l}
\text{~~CH}_2\text{--}\overset{\overset{\displaystyle CH_3}{|}}{\underset{\underset{\displaystyle R}{|}}{C}} + HS\text{--}\overset{}{\underset{\underset{\displaystyle CH_2\text{--}COOH}{|}}{CH}}\text{--}COOH \longrightarrow \text{~~CH}_2\text{--}\overset{\overset{\displaystyle CH_3}{|}}{\underset{\underset{\displaystyle R}{|}}{CH}} + S\text{--}\overset{}{\underset{\underset{\displaystyle CH_2\text{--}COOH}{|}}{CH}}\text{--}COOH
\end{array}
$$

$$
HOOC\text{--}\overset{}{\underset{\underset{\displaystyle HOOC\text{--}CH_2}{|}}{CH}}\text{--}S + nCH_2\text{=}\overset{}{\underset{\underset{\displaystyle R}{|}}{CH}} \longrightarrow HOOC\text{--}\overset{}{\underset{\underset{\displaystyle HOOC\text{--}CH_2}{|}}{CH}}\text{--}S\text{--}CH_2\text{--}\overset{}{\underset{\underset{\displaystyle R}{|}}{CH}}\text{--}CH_2\text{--}\overset{}{\underset{\underset{\displaystyle R}{|}}{CH}}\text{~~}
$$

Another example [106] makes use of α-thioglycerol as a transfer agent, and the macromonomers obtained exhibit the following structure:

$$
\overset{}{\underset{\underset{\displaystyle COOCH_3}{|}}{CH}}\text{--}CH_2\text{--}(\overset{\overset{\displaystyle CH_3}{|}}{\underset{\underset{\displaystyle COOCH_3}{|}}{C}}\text{--}CH_2)_{n-1}\text{---}S\text{--}CH_2\text{--}\overset{}{\underset{\underset{\displaystyle OH}{|}}{CH}}\text{--}\overset{}{\underset{\underset{\displaystyle OH}{|}}{CH_2}}
$$

They cover a molecular weight range of 3000 to 6000 $g \cdot mol^{-1}$ and have been adequately characterized. The molecular weights determined by GPC and by OH titration agree very well, and the polydispersity ratio does not exceed 1.6 which is quite satisfactory.

These species can be used in stepwise polymerization processes. Each macromonomer bearing two terminal vicinal carboxy or hydroxy groups yield a graft.

2.5.2.3 Telomerization

A special case of free-radical transfer reactions is that of telomerization. Telomerization has gained great attention in recent years, especially for monomers such as vinyl chloride, vinylidene chloride, and fluorine-containing monomers. They often enable the control of the molecular weights of the formed species and yield chains bearing groups originating from the telogen.

Telomerization can be initiated either by free-radical initiators such as benzoyl peroxide or azo-bis-isobutyronitrile (AIBN) or by redox processes first investigated by Asher and Vofsi [107], and developed more recently by Pietrasanta and co-workers [108]. Redox systems generally consist of a metal salt (Fe^{+++} or Cu^{++}) and a reducing agent such as benzoin. Telogens include CCl_4 or CCl_3Br and trichloromethyl derivatives such as CCl_3—R some of which are themselves monoadducts obtained by telomerization between CCl_4 and an unsaturated compound [109]. The degree of polymerization of the obtained telomers largely depends on the system chosen and the experimental conditions. Free radical and redox systems differ very much in this respect [110]. Using redox systems, it is generally easier to obtain adducts or well-defined oligomers which can be separated by fractional distillation.

A typical case in which a 1:1 adduct is formed quantitatively is that of allylic derivatives [109] (allyl alcohol or allyl acetate):

$$CCl_4 + CH_2 = CH - CH_2OH \rightarrow CCl_3 - CH_2 - CHCl - CH_2OH$$

Using free-radical initiation, macromolecules with much higher molecular weights are obtained. The molecular weights may sometime be controlled by proper choice of the molar ratio of monomer to telogen and of the reaction temperature. Telomerizations have been used by Pietrasanta, Boutevin [111] and their coworkers to produce macromonomers by a two-step process. Thus, ω-functional polymers have been prepared by telomerization with functional telogens and subsequent reaction of the terminal functions with an unsaturated reagent. In the following some interesting examples are described.

Chlorotrifluoroethylene macromonomers [112] have been obtained as follows: the monomer is reacted in acetonitrile with carbon tetrachloride in the presence of the Fe^{+++}/benzoin redox system whereby oligomers with polymerization degrees ranging from 1 to 20 are obtained:

$$CCl_4 + CF_2 = CFCl \xrightarrow{F_e^{+++}/benzoine} CCl_3-(CF_2-CFCl)_n-Cl \qquad (I)$$

Species of type (I) can be used as a telogen with allyl acetate or allyl alcohol whereby monoadducts are formed:

$$(I) + CH_2 = CH-CH_2OH \rightarrow$$
$$HO-CH_2-CHCl-CH_2-CCl_2-(CF_2-CFCl)_n-Cl$$

To fit these ω-hydroxy telomers with terminal double bonds they were esterified with acrylic acid in the presence of p-toluenesulfonic acid. Subsequent azeotropic distillation of the formed water yields macromonomers exhibiting the following structure

$$CH_2 = CH-COO-CH_2-CHCl-CH_2-CCl_2-(CF_2-CFCl)_nCl$$

These macromonomers were successfully copolymerized with methyl methylacrylate and vinyl acetate:

Vinyl chloride was telomerized both with 2-hydroxyethyl trichloroacetate and 2,4,4,4-tetrachlorobutanol (the latter compound is the monoadduct of allyl alcohol and CCl_4) using redox initiation [111]. Telomers bearing at chain end a hydroxy group were obtained:

$$HO-CH_2-CH_2-O-CO-CCl_2-(CH_2-CHCl)_n-Cl$$

$$HO-CH_2-CHCl-CH_2-CCl_2-(CH_2-CHCl)_n-Cl$$

These species can be subsequently reacted either with acrylic (or methacrylic) acid as described or with methacryloyl chloride in the presence of lauryldiethylamine to remove HCl and thus to prevent their addition to the terminal double bond which would occur in the absence of the amine.

— Vinyl chloride can also be telomerized by free-radical initiation [110] in the presence of trichloroacetic acid or of the monoadduct of CCl_4 and acrylic (or methacrylic) acid (obtained with Cu^{++}/CH_3CN):

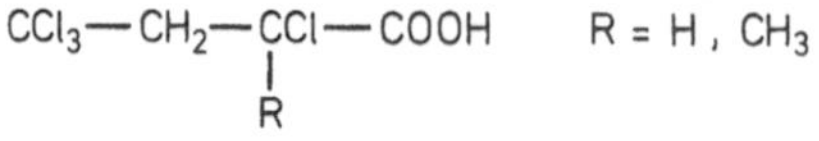

$$CCl_3-CH_2-\underset{\underset{R}{|}}{C}Cl-COOH \qquad R = H, CH_3$$

The same monomer also yields telomers with thioglycolic acid [111] and monothioglycol using free-radical initiation (AIBN) under appropriate conditions. The authors synthesized macromonomers by reaction of glycidyl methacrylate with the ω-carboxy telomers

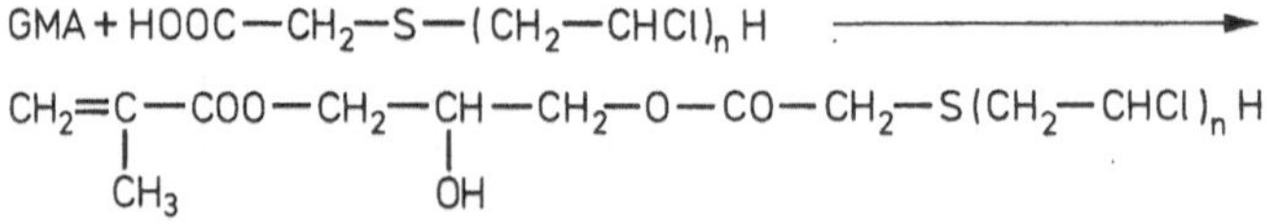

$$GMA + HOOC-CH_2-S-(CH_2-CHCl)_n H \longrightarrow$$

$$CH_2=\underset{\underset{CH_3}{|}}{C}-COO-CH_2-\underset{\underset{OH}{|}}{CH}-CH_2-O-CO-CH_2-S(CH_2-CHCl)_n H$$

and by reaction of methacryloyl chloride with ω-hydroxy telomers in the presence of lauryldiethylamine:

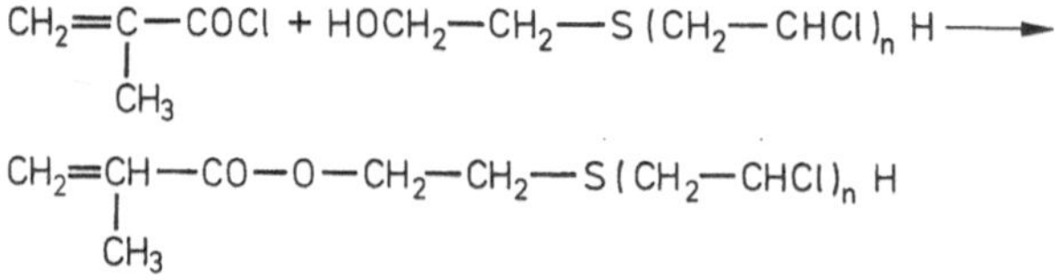

$$CH_2=\underset{\underset{CH_3}{|}}{C}-COCl + HOCH_2-CH_2-S(CH_2-CHCl)_n H \longrightarrow$$

$$CH_2=CH-CO-O-CH_2-CH_2-S(CH_2-CHCl)_n H$$
$$\underset{CH_3}{|}$$

The macromonomers thus obtained exhibit molecular weights as low as 1000. They were copolymerized with monomers such as styrene and butyl acrylate whereby graft copolymers with poly(vinyl chloride) grafts were obtained.

3 Polymerization and Copolymerization of Macromonomers

In this section homopolymerization of macromonomers and their copolymerization with vinylic or acrylic monomers will be reviewed.

3.1 Homopolymerization of Macromonomers

It could be expected that the rate of polymerization of macromonomers is influenced by the low concentration of the active sites in a macromonomer solution and possibly also because of the lower accessibility of the growing radicals and of the terminal unsaturated bonds of the macromonomer.

This point of view is however questioned by Yamashita [98] and by Asami [21] whose results seem to indicate that the reactivity of a double bond does not depend on the length of the chain, to which it is attached. However, they usually obtained low yields of the products resulting from the polymerization of macromonomers and the degrees of polymerization were also relatively low as far as one can juge from the reported characterization data.

Another factor has to be considered. When the macromonomer chain segments can give rise to transfer reactions, the probability of such events is enhanced because the number of chain segments per unsaturation is high. Homopolymerization of such macromonomers should be expected to involve additional branching and possibly crosslinking. A typical example is that of PEO macromonomers because oxyethylene units are known to induce transfer reactions [113].

Only few free-radical homopolymerizaions of macromonomers have been reported so far in spite of the importance of such highly branched compact macromolecules exhibiting very high segment densities in the vicinity of the backbone chain. Before giving some typical examples of macromonomer homopolymerization, attention should be drawn to two specific difficulties involved, namely separation and characterization:

— When a macrom nomer is polymerized, the determination of the conversion requires a quant.tative separation of the unreacted macromonomer which is in some cases rather difficult and requires fractionation. GPC can be very helpful to evaluate the amount of unreacted macromonomer and to check the accuracy of the fractionation.

The characterization of macromonomer homopolymers (or even copolymers) by GPC alone may lead to erroneous conclusions.

The hydrodynamic volume (which is the parameter that determines the chromatographic retention in GPC [114]) of such highly branched macromolecules is very small as compared with that of linear homologues of the same molecular weight. Therefore, molecular weights calculated from the GPC diagrams and based on calibration curves established with linear homologues are obviously underestimated. Even the universal calibration [114] that takes account of limited branching should be used with great caution in characterizing compact polymer molecules as those originating from macromonomer homopolymerization.

3.1.1 Free-Radical Homopolymerization of Macromonomers

Polyoxyethylene macromonomers bearing terminal methacryloyl groups were homopolymerized by means of free-radical initiators in the presence of an additional transfer agent to keep the degree of polymerization low (and possibly to lower the probability of transfer to PEO chains that could induce cross-linking). This procedure developed by Watanabe [62] led to what he refers to as "curtain polymers" the characterization of which by GPC is, however, misleading for the reasons discussed above. Gramain and Frere [34] also studied the free-radical homopolymerization of ω-methacryloyl PEO macromonomers; in aqueous solution and with potassium persulfate as the initiator they obtained soluble polymacromonomers containing up to 20 macronomer units.

ω-(4-vinylbenzyloxy)-poly-THF macromonomers have also been homopolymerized by free-radical initiators such as AIBN. Asami [50, 115] obtained yields of 50–70% after 20 hours at 60 °C but the molecular weights were reported to be rather low. Here again the degrees of polymerization were calculated from the GPC data. However, without any absolute molecular weight determination definite conclusions cannot be drawn. Even the polydispersity ratio must be underestimated because the hydrodynamic volume of such compact molecules should not markedly depend on the number of branches [116].

Polystyrene macromonomers bearing terminal methacryloyl groups have been polymerized in benzene solution using AIBN as the initiator [24]. The yields measured seem to indicate that the longer the macromonomer chain, the lower the rate of growth. However, the polymerization degrees measured by light scattering range from 20 to 40 whereas the apparent DP's determined from GPC data are of the order of 10. This clearly evidences the highly branched character of these "polymacromonomers".

The homopolymerization of ω-(4-vinylbenzyl)polystyrene macromonomers was also investigated kinetically by Asami [21] under quite different conditions, namely very high amounts of initiator and high overall concentrations. Thus, the molecular weights (even if underestimated) are very low. Under these conditions, the rate of polymerization does not depend on the length of the side chains. However, these particular conditions which favour initiation and termination processes cannot be illustrative of regular polymerizations.

3.1.2 Anionic Homopolymerization

Only few attempts to homopolymerize macromonomers anionically have been reported. It should be pointed out that obviously the terminal unsaturated group should be susceptible to attack by carbanionic sites (which limits the number of candidates) and that the macromonomer chain should contain neither proton-donating groups nor highly electrophilic functions capable of deactivating the living sites. Owing to the higher stability of the carbanionic sites there are greater chances of obtaining the expected homopolymers, even though the polymerization rates might be smaller for the reasons quoted above.

The anionic homopolymerization of polystyrene macromonomers was carried out successfully. The methacrylic ester sites at the chain end do not require very strong nucleophiles to be initiated: diphenylmethylpotassium was used, and the process was carried out at − 70 °C in THF solution [24]. The products are comparable with those obtained by free-radical polymerization. The molecular weight distribution should be narrower but this cannot be easily checked because these polymer species are highly branched and compact as already mentioned.

With polyoxyethylene macromonomers carrying either a methacrylic ester function or an α-methylstyryl group at the chain end, the anionic polymerization should be carried out at low temperature to avoid side reactions involving the ester function or to shift the propagation — depropagation equilibrium, respectively. Here, an experimental difficulty arises which could not yet be overcome. At temperatures below 0 °C in solvents such as THF or dimethoxyethane, (and even more so in benzene) crystallization of the PEO occurs, and the macromonomer precipitates, thus preventing polymerization. The likelihood of side reactions is even increased by the fact that alkali metal cations are very well solvated by the PEO chains, thus increasing the reactivity of the carbanions. Therefore, this reaction cannot be successfully carried out at room temperature with macromonomers bearing methacrylic ester functions at the chain end.

Anionic homopolymerization of poly-THF macromonomers bearing terminal styryl, α-methylstyryl or methacryloyl groups has not been reported so far.

3.1.3 Copolymerization of two Macromonomers

Copolymerization of two different macromonomers (namely ω-styrylpolystyrene and ω-styryl-poly-THF) has also been attempted by Asami [50] in benzene solution at 60 °C for 48 hours, using free-radical initiation. Both macromonomers take part in the process, and the formation of the "curtain copolymer" was evidenced by the GPC diagrams. However, no precise characterization has been given, and no information is available on the sequence distribution of the side chains which are incompatible with each other. The morphology of such highly branched heterograft copolymers should be of great interest because they exhibit *both* a high segment concentration *and* high tendency towards intramolecular phase separation.

3.2 Copolymerization of Macromonomers with Comonomers

Free-radical copolymerization of a macromonomer with a vinylic or acrylic comonomer has been and is still the major field of application of macromonomers since it provides easy access to graft polymers. A large number of patents and papers deal with this problem, and even though the characterization of the graft copolymers is often far from being adequate, this method has proved successful in numerous applications.

3.2.1 Fundamentals

The copolymerization of macromonomer with comonomer is governed by the general rules of copolymerization, the ability of any of the two polymerizable species present to participate in the process being determined by the radical reactivity ratios r. Let us denote the macromonomer as M and the comonomer as A. The well-known instantaneous composition law applies to the copolymer formed:

$$\frac{d[A]}{d[M]} = \frac{[A]}{[M]} \frac{r_a[A] + [M]}{r_m[M] + [A]} .$$

[A] and [M] are the concentrations of A and M respectively at time t, d[A] and d[M] the amounts of A and M consumed during time interval (t, t + dt), and $r_a = k_{aa}/k_{am}$ where k_{aa} and k_{am} are the rate constants of the addition of A and M respectively to the radical A.

In most copolymerization experiments the molar macromonomer concentration [M] is low with respect to the molar concentration [A].

Consequently, the above equation reduces to [117] [3]

$$\frac{d[A]}{d[M]} = r_a \frac{[A]}{[M]} \qquad \text{if } [A] \gg [M]$$

which can be integrated to [118]:

$$r_a = \frac{\ln [A]_0/[A]}{\ln [M]_0/[M]} = \frac{\ln (1 - x_a)}{\ln (1 - x_m)}$$

---

3 A similar situation is encountered in so-called "ideal" copolymerizations, characterized by the relation $r_a \times r_m = 1$, but in this case the above equations are valid over the entire composition range

where x_a and x_m are the fractional conversions of A and M at time t and $[A]_0$ and $[M]_0$ the initial molar concentration of A and M in the reaction medium. The above equations imply that the actual value of r_m is of no importance to the process, provided the molar amount of the macromonomer used is small.

Hence, in macromonomer copolymerization the ratio of the molar amounts of A and of M incorporated instantaneously into the copolymer is proportional to the ratio of their molar amounts in the feed. This does not prevent compositional heterogeneity in samples obtained in the case of high conversions; however, the amplitude of the fluctuations in the composition generally remains acceptable for conversions up to 50–70%, because of the low molar concentrations of M usually chosen.

If $r_a > 1$, the consumption of monomer A is faster than that of the macromonomer. Therefore, as conversion increases the proportion of M in the feed tends to increase; consequently, the proportion of grafts incorporated in the copolymer also increases.

If $r_a < 1$, the opposite is true: A is consumed at a lower rate than M, which implies that the percentual amount of M in the feed tends to decrease as conversion progresses.

The probability of finding an AA diad at any given place along the chain can be expressed as

$$p_{aa} = F_a \cdot \frac{k_{aa}[A]}{k_{aa}[A] + k_{am}[M]} = F_a \cdot \frac{r_a f_a}{r_a f_a + f_m} = F_a^2$$

Here f_a and f_m are the molar proportions of A and M in the monomer mixture, and F_a is the mole percent of A units in the copolymer formed instantaneously. Similarly, the probability of finding AM and MA diads at any given place along the chain is expressed as

$$p_{am} = p_{ma} = F_a \frac{f_m}{r_a f_a + f_m} = F_m \frac{r_a f_a}{r_a f_a + f_m} = F_a \cdot F_m$$

The probability of finding MM diads along the chain is negligible if f_m remains below 0.05:

$$p_{mm} = F_m^2 = \frac{f_m^2}{(r_a f_a + f_m)^2} \simeq 0$$

From these expressions it follows that the instantaneous distribution of the M units along the backbone (and consequently that of the grafts) formed instanteneously follows the Bernouillian statistics, provided the molar amount of the macromonomer in the reaction mixture is small enough. Fluctuations in composition generally remain within rather narrow limits, provided the conversion is not too high; consequently, the content of grafts in a graft copolymer sample remains within narrow limits, too.

Revillon and Hamaide [119] studied the kinetics of the copolymerization of butyl acrylate (BA) with an ω-styrylpolyoxyethylene macromonomer by means of high-resolution GPC to characterize the amounts of graft copolymer formed and of the macromonomer and comonomer consumed. In this case, the rather low molecular weight of the macromonomer ($200 \text{ g} \cdot \text{mol}^{-1}$) allowed accurate determinations over a wide range of initial compositions. From the obtained data the reactivity ratios of the two monomers were determined according to the method of Kelen-Tüdös [120]. A good agreement was found between the experimental and the calculated diagrams of instantaneous composition. In this case, $r_{BA} = 1.5$ and $r_m = 0.2 \pm 0.1$; the lower accuracy of the r_m value is not a disadvantage if the molar emount of the introduced macromonomer is low, as discussed earlier.

Another interesting investigation was carried out by Yamashita et al. [98]. Copolymerization of ω-methacryloylstearyl methacrylate macromonomers with methyl methacrylate was studied over a wide range of compositions, and it was found that all copolymers formed (up to conversions of 60 %) exhibit the same composition as the initial macromonomer-comonomer mixture. It thus seems that for this system (in with both species are methacrylic esters) r_a and r_m should be very close to unity, thus preventing fluctuations in the composition and yielding a strictly random distribution of grafts along the backbone chain.

3.2.2 Survey of Graft Copolymer Synthesis using Macromonomers

For the last 20 years anionic polymerization techniques have enabled to synthesize well-defined graft copolymers using either carbanionic deactivation or carbanionic initiation. The former method implies that a living polymer precursor is reacted onto a polymer backbone carrying electrophilic functions. Typical examples [121] include grafting of polystyrene onto a poly(methyl methacrylate) backbone [122] or grafting of poly(oxyethylene) chains (bearing *one* alkoxide function) onto a polystyrene backbone that had been partially chloromethylated [123]. The latter method consists in partial metalation of a polymer backbone, and in using these sites to initiate the polymerization of an adequate monomer [124].

Even though some techniques of cationic grafting [125] and of free-radical grafting[4] have been developed [126, 127], the number of systems that can actually be used is very small, and the search for new and more general methods, applicable to systems that cannot respond to ionic techniques, has been one of the determining factors for research in the field of macromonomers.

The chief reason for the interest in graft copolymers originates from the incompatibility between polymer chains of different chemical nature. Intramolecular phase separation results, because grafts and backbone repell each other, and these compounds exhibit a marked tendency to form mesomorphic phases like block copolymers and soaps do. When these species are mixed with a solvent that exhibits a preferential affinity for one of the components (grafts or backbone) the incompatibility may be enhanced. This intramolecular phase separation has led to a number of applications. If small amounts of a graft copolymer are included into a homopolymer of the same nature as the grafts (or the backbone), surface modifications can result as described below.

---

4 Grafting can also be achieved by means of high energy irradiation techniques. However, in many cases, side reactions such as cross-linking, fragmentation, and homopolymerization may also occur

Special attention was granted to amphiphilic graft copolymers in which hydrophilic grafts are attached to a hydrophobic backbone (or vice versa). Such materials can be used as emulsifiers, compatibilizers, surface modifiers, adhesives, etc. and they may play a role in many industrial developments.

The most representative research projects concerning graft polymer synthesis by means of macromonomer copolymerization will be reviewed in this section, whenever some kind of characterization of the compounds obtained is presented. Indications on potential applications of these materials will be given whenever available.

3.2.2.1 Poly(oxyethylene)Macromonomers

In a patent dated 1965 Stowe [35] laid the basis for the copolymerization of PEO macromonomer with comonomers such as acrylonitrile. It was searched for an increased wettability of polyacrylonitrile films or fibers by a "permanent" surface modification. ω-Styryl poly(oxyethylene) macromonomers readily copolymerize with acrylonitrile in water emulsions. They can also be copolymerized with styrene-sulfonates in the presence of poly(vinylpyrrolidone). The presence of small amounts of such copolymers in polyacrylonitrile fibers was shown to increase their wettability and their receptivity to dyes and to make them more resistant to electric loading (antistatic fibers). No characterization data on the copolymers formed have been reported.

Hamaide and Guyot [36] copolymerized ω-styryl polyoxyethylene macromonomers with styrene in toluene solution using AIBN as the initiator. They also produced grafted networks by the addition of 20–40 wt-% of divinylbenzene to the above reaction mixture. In all these experiments, the molar amount of the PEO macro-monomer was of the order of 5%. The purpose of these synthesis was to examine the ability of the PEO grafts to solvate cations, and thus to accelerate reactions involving ion-pair dissociation. Indeed, these polymers and networks bearing PEO side chains exhibit good complexing properties towards alkaline metal cations, as revealed by kinetic measurements on the Williamson condensation reaction between potassium phenoxide and alkyl bromides. Potassium phenoxide is insoluble in toluene, and in this medium the reaction does not proceed. However if a small amount of the above copolymer or network is added, the reaction proceeds to high conversions. The soluble polymers were found to be more efficient than the swollen networks containing the same molar amount of PEO grafts.

Similar studies were performed in our laboratory [128]: ω-Methacryloyl-PEO (or α,ω-bis-α-methylstyryl-PEO) was copolymerized both with styrene and alkyl methacrylates, the molar amount of the macromonomer being of the order of 5% (which nearly corresponds to 50% by weight of a macromonomer with the molecular weight 2000). The reaction was carried out in benzene solution with AIBN as the initiator. Isolating the graft copolymer and making it free from unreacted macro-monomer (the yields were kept below 70%) were rather difficult because of the amphiphilic properties of the copolymers. If they contain more than 35–40 wt-% of PEO they give rise to very stable emulsions in methanol and even in water. The most efficient method of isolating the copolymer involves precipitation from its benzene solution by the addition into the nonsolvent heptane. Fractionations were carried out with the same system. In some cases, removal of the unpolymerized macromonomer can be performed by washing the dried raw sample either with methanol or cold water. The copolymers were characterized, and only slight fluctuations in the composition were found [129]. The amphiphilic properties of these copolymers have been investigated.

Further studies on the preparation of networks coated with hydrophilic chains at the surface are being made.

Watanabe [62] studied systematically the copolymerization of ω-methacryloyl-polyoxyethylenes, with monomers such as acrylonitrile, styrene, butyl methacrylate, and methacrylic acid. It should be mentioned that the macromonomers that he prepared are very short so that no difficulties were encountered to isolate the graft copolymers formed. There are many applications for these graft copolymers, e.g. as additives in polyacrylonitrile films and fibers they cause improved antistatic properties. They have been tested as varnishes, coatings, and wood dimensional stabilization agents.

PEO macromonomers were also reacted with isocyanates to introduce urethane linkage, the methacrylic ester functions being used to cure the material in a later step.

Very recently Saegusa [17] investigated the cationic copolymerization of ω-oxazolyl-polyoxy ethylene macromonomers with 2-phenyloxazoline as the comonomer. The copolymerization was carried out in acetonitrile at 80 °C using BF_3/Et_2O as the initiator. The copolymer yields were of the order of 60–80%. The amount of macromonomer incorporated was slightly below its amount in the reaction mixture and the molecular weights were rather low (5 to 15000 g/mol), as determined by vapor pressure osmometry. However, fractionation of the copolymer showed that relatively large fluctuations in molecular weight and in composition occur within the sample.

This very interesting investigation is to our knowledge the first exemple of a macromonomer fitted at the chain end with a heterocycle susceptible to cationic homo- and copolymerization.

3.2.2.2 Poly-THF Macromonomers

The copolymerization of ω-methacryloyl-polytetrahydrofuran with comonomers such as butyl methacrylate (BuMA) or styrene using AIBN as the initiator was investigated by Sierra-Vargas et al. [117]. With styrene, the macromonomer content of the graft copolymer is slightly higher than its content in the reaction mixture; with butyl methacrylate the reverse is true. This result implies that the reactivity ratio of BuMA in this system is greater than unity, in contrast to Yamashita's result [98]. Fractionations carried out on the graft copolymer samples — after removal of the unreacted macromonomer — evidenced that fluctuations in the composition are low.

Similar work was performed by Asami on the copolymerization of poly-THF macromonomers: The formation of graft copolymers was evidenced though their molecular weights, determined by GPC, were obviously underestimated.

It should be pointed out that the same type of graft copolymers has been obtained by Asami [130] in a quite different manner involving deactivation of living poly-THF on a backbone chain of polystyrene with some randomly distributed p-hydroxystyrene units which have been metalated beforehand.

3.2.2.3 Polystyrene Macromonomers

Waak [14] successfully copolymerized ω-allylpolystyrene macromonomers with acrylonitrile and several other monomers. Graft copolymers resulted in spite of the fact that allylic double bonds are not very suitable for free-radical copolymerization. However, a detailed characterization of these copolymers has not been given.

Milkovich and his coworkers [18, 131] used well defined anionically prepared ω-methacryloylpolystyrene macromonomers with molecular weights in the range from 5000 to 20 000 g · mol^{-1} for free-radical copolymerization experiments with monomers such as methyl acrylate, butyl acrylate, vinyl chloride, acrylonitrile. Solution polymerization, aqueous suspension polymerization (producing beads) and emulsion polymerization (yielding stable latices) were investigated. The conversion of the macromonomers was followed by GPC and it was found that when the comonomer conversion reaches 80 % unreacted macromonomer can no longer be extracted. In view of the fact that the weight fraction of the macromonomer generally ranges from 30 to 50 % (its mole fraction being very small because of its relatively high molecular weight: $\simeq 1$ %), the assumption of a random distribution of the grafts along the backbone chain appears to be justified.

Samples with only few grafts exhibit the typical properties of thermoplastic elastomers, originating from intramolecular phase separation; physical cross-linking is attributed to the fact that polystyrene grafts belonging to different polymer molecules are located in a common microdomain. When the polystyrene content is higher, the materials are strong, flexible thermoplastics exhibiting satisfactory mechanical properties. Copolymers of polystyrene macromonomers with acrylonitrile or vinyl chloride are transparent and more easily processed than the PAN or PVC homopolymers.

So far, no data on the molecular weights of these graft copolymers have been reported nor on their compositional heterogeneity.

Milkovich [18] et al. also investigated the copolymerization of well-defined polystyrene (or polydiene) macromonomers with monomers such as ethylene, propene, and isobutene; emphasis was laid on Ziegler-Natta type copolymerization with olefins and cationic copolymerization with isobutene. Also, free-radical processes were used for the copolymerization of the macromonomers with acrylamide, vinyl acetate, and those already quoted. So far, only few characterization data have been reported on the graft copolymers themselves. The macromonomers have in most cases quite long chains (M_n from 5000 to 50 000 g · mol^{-1}) and even though the number of the grafts is generally low, they nevertheless constitute 30–80 % by weight of the described materials.

Yamashita et al. [101] studied the copolymerization of ω-methacryloylpolystyrene macromonomers with various comonomers. These macromonomers exhibit molecular weights around 5000 g · mol^{-1}, and they were copolymerized with either 2-hydroxyethyl methacrylate or perfluoroalkyl acrylates $CH_2 = CH—COO—CH_2—CH_2C_nF_{2n+1}$ ($n \cong 9$). The first system yields amphiphilic graft copolymers, because poly-HEMA is hydrophilic and the polystyrene grafts are hydrophobic. The copolymerizations were carried out to high conversions (> 80 %) with molar amounts of the macromonomer up to 6 %; the fluctuations in composition seem to be rather high since a large fraction of HEMA-rich copolymer is present in the samples. A better homogeneity is attained when the polystyrene macromonomer is copolymerized with a mixture containing equal fractions of HEMA and MMA.

The second system investigated [101] (polystyrene macromonomer and perfluoro-alkyl acrylate) is also of great interest. The polymerization is carried out in trifluoro-benzene with AIBN as the initiator to a conversion of the order of 60%. The graft copolymer formed is soluble in a number of solvents in which the poly(perfluoro-alkyl acrylate) backbone would be insoluble, e.g. in THF and diethyl ether. The easy formation of foams indicates the low surface energy which is characteristic of fluorinated polymers. Double-detection GPC (UV and refractive index) showed that the distribution of polystyrene branches within the sample was quite uniform.

3.2.2.4 Poly(alkyl Methacrylate) Macromonomers

Yamashita [2, 99, 132] studied the copolymerization of ω-methacryloyl-poly(methyl methacrylate) macromonomers with 2-hydroxyethyl methacrylate and the perfluorin-ated alkyl acrylate mentioned above. He also terpolymerized the macromonomer with either of these two comonomers and methyl methacrylate. The molecular weights of the PMMA macromonomers ranged from 3000 to 9000. The yields were of the order of 70% and the macromonomer content of the graft copolymers was generally slightly lower than that of the initial mixture. The molecular weights determined by GPC do not provide valuable information. The interesting feature of Yamashita's work is that it proved that ample surface modifications can be performed on PMMA films containing amounts of these graft copolymers. The PMMA grafts play the role of anchoring segments as they are compatible with the PMMA, and the backbone chains are preferentially located at the film surface. Compared with unmodified PMMA films, the material containing a small amount of poly-HEMA graft copolymer is more hydrophilic, and the material containing poly(perfluoroalkyl acrylate) graft copolymer is far more hydrophobic. These results have been obtained by measurements of the contact angle of water droplets on the film surface and by detailed investigation of the polymer surface by ESCA and by ATR-IR. The "surface accumulation" [133] of both HEMA and perfluoroalkyl acrylate segments is easily detected, even for very low graft copolymer contents in the PMMA films.

Yamashita et al. [2, 99] also investigated the copolymerization of PMMA macro-monomers both with a mixture of HEMA and perfluoroalkyl acrylate and with a MMA-methacrylic acid mixture; here again, the PMMA grafts originating from the macromonomer play the role of anchoring segments, and surface accumulation of the functional backbone segments is well established.

The same authors [132] studied the copolymerization of short ω-methacryloyl-PMMA macromonomers with monomers such as styrene, vinyl acetate and acrylo-nitrile using AIBN as the initiator. They found that the copolymers formed are free of homopolymer; these graft copolymers possessing a small number of very short grafts exhibit properties similar to those of the backbone chains.

Waite [93, 100] investigated the copolymerization of lauryl methacrylate macromono-mers with methyl methacrylate and used the graft copolymers as stabilizers for the dispersions. The polar backbone is more compatible with dispersed polar droplets and the grafts, which are much less polar, are miscible with the continous phase, thus yielding stable "oil-in-oil"-dispersions.

Stearyl and lauryl methacrylate macromonomers were also synthesized by Ito and Yamashita [98] using the free-radical technique already described. These macromonomers having molecular weights in the range 2000–5000 g · mol^{-1} were copolymerized with methyl methacrylate to 50–70% conversion. However, no accurate molecular weights of the graft copolymers have been reported so far but their composition is very close to that of the feed from which the authors concluded that the reactivity ratios are close to unity. To study the occurrence of phase separation in these graft copolymers they were used as the stationary phase in inverse gas chromatography [134]. The chromatographic retention of dodecane, a non-polar hydrocarbon, clearly indicates microphase separation, the stearyl methacrylate segments constituting the continuous phase down to about 30% by weight of stearyl methacrylate which means 1 to 2 branches per 100 units MMA., depending on the molecular weight of the macromonomer.

This brings new evidence of the "surface accumulation" of one of the constituents, a property that could lead to a great variety of applications, whenever surface modification is needed, e.g. in films, fibers, adhesives, coatings, varnishes, affinity chromatography etc.

3.2.2.5 Polyisobutene Macromonomers

Kennedy [67, 77, 118] studied the ability of ω-styryl-polyisobutene macromonomers to undergo free-radical copolymerization with either styrene or butyl or methyl methacrylate. Here, the macromonomers exhibited a relatively high molecular weight of 9000, and the reaction was stopped after roughly 20% of the comonomer had been converted. The radical reactivity ratios of styrene and methyl methacrylate with respect to macromonomer were found to be equal to 2 and to 0.5, respectively. From these results, Kennedy concluded that in the ω-styrylpolyisobutene/styrene system the reactivity of the macromonomer double bond is reduced whereas with methacrylate as the comonomer the polar effect is the main driving force, yielding reactivities similar to those observed in the classical system styrene/MMA.

However, no indications are given of the molecular weights nor of the compositional heterogeneity of the graft copolymers, their properties or potential applications. The existence of a microphase separation was evidenced however on methyl methacrylate/polyisobutene graft copolymers by means of DSC measurements. Two glass transitions were observed at −68 °C (polyisobutene) and at 102 °C (PMMA), i.e. at the same temperatures at which they occur in the corresponding honopolymers.

The polyisobutene macromonomers bearing a methacryloyl group at both chain ends have not been copolymerized so far. Yet this would provide a route to a new kind of tailor-made networks.

3.2.2.6 Polydimethylsiloxane Macromonomers

The ω-styryl- and ω-methacryloylpolydimethylsiloxane macromonomers synthesized by Yamashita [29] were copolymerized with styrene and methyl methacrylate, respectively, using AIBN as the initiator. The yields were about 80%, and the unreacted macromonomer was extracted. There is a good agreement between the molar amount of the macromonomer in the reaction mixture and the number of grafts in the copolymer. This result was considered as an argument in favor of reactivity ratios close to unity, in other words, the reactivity of an ω-styrylmacromonomer and that of styrene itself should be the same. However, this conclusion is not entirely convincing because of the very high yields attained. The molecular weight of the graft copolymers were determined by membrane osmometry and GPC, and the difference between the two values clearly illustrates the highly branched structure of the graft copolymers obtained.

No indication is given of potential applications of these graft copolymers. Such materials could obviously be of interest for biomedical applications owing to the good biocompatibility of PDMS polymers.

3.2.2.7 Polyamine Macromonomers

The polyamine macromonomers developed by Tsuruta and his coworkers [81, 84, 135] were also copolymerized with styrene, using AIBN as the free-radical initiator.

The macromonomers obtained by self-condensation of the monoadduct originating from p-divinylbenzene and N,N'-diethylethylenediamine have molecular weights ranging from 3000 to 13000. The copolymers formed to conversion of the order of 20–40% can be easily liberated from unreacted polyamine macromonomer. The molar amount of the macromonomer in the initial mixture ranges from 0.5 to 4%. The graft content of the copolymers is slightly lower but the molecular weights obtained are rather high.

Polyamine macromonomers containing crown ethers [84] exhibit molecular weights from 1000 to 4000, i.e. degrees of polymerization ranging from 2.5 to 10. These compounds are soluble in many solvents including water but insoluble in ether, acetone and DMSO. Thus, after copolymerizing these macromonomers with styrene to conversion of about 20%, the unreacted macromonomer is easy to remove. The copolymers have rather low molecular weights but they contain grafts (in proportions close to those of the macromonomer in the reaction mixture), and it can be anticipated that these species could be of major interest for cation binding purposes.

The p-DVB-piperazine adducts can also undergo self condensation whereby the macromonomers formed exhibit rather low molecular weights, and are insoluble in the reaction mixture (benzene or THF). They dissolve only upon the addition of an acid or in hot chloroform. Free-radical copolymerization of this macromonomer with styrene was carried out in benzene in the presence of some acetic acid (to obtain a homogeneous reaction mixture) to yields of about 20% [84]. Here again separation of the unreacted macromonomer is possible, and the polyamine content of the graft copolymer is very close to the amount contained in the reaction mixture.

In all three cases, it can be assumed that the grafts are distributed at random along the polystyrene backbone chain. These species exhibit a strong tendency to intramolecular phase separation as it could be expected with backbones and grafts of different chemical nature. Moreover, they display quite different affinities to solvents. Applications of these compounds are searched for as materials of biomedical interest.

3.2.2.8 Polypeptide Macromonomers

As mentioned earlier Maeda and Tsuruta [87, 88] developed a method to synthesize ω-styrylpolypeptide macromonomers: The polymerization of NCA derived from L-benzyl glutamate is initiated by the primary amine functions of the monoadduct of DVB and N-methylethylenediamine. The obtained degrees of polymerization are close to 10. Copolymerization of styrene or methyl methacrylate with these polypeptide macromonomers (molar amount 0.5 to 2% in the initial mixture) was carried out in benzene, using AIBN as the initiator. The process yielded the expected graft copolymers with a proportion of grafts very close to that calculated assuming equal reactivities of the double bonds. The unreacted macromonomer is quantitatively extracted with trifluoroacetic acid. Circular dichroism measurements showed that neither racemization nor conformational changes did occur in the polypeptide grafts upon copolymerization. Transmission electron microscopy clearly revealed that here again extensive phase separation occurs. Such materials could be of interest for biomedical applications. However, no details have been reported.

The same type of graft copolymers with a polystyrene backbone and polypeptide grafts has been synthesized by a different route [89]: The monoadduct of DVB and N-ethylethylenediamine was copolymerized with styrene, yielding a random copolymer with pendent primary amine functions. The latter were used as initiators for the subsequent polymerization of the NCA derived from benzyl L-glutamate.

3.2.2.9 Macromonomers for Step Growth Polymerization

ω-Dihydroxypoly(methyl methacrylate) [106] was mixed with another diol (1,2-hexanediol) and reacted with a stoichiometric amount of a diisocyanate (TDI or HMDI) in N-methylpyrrolidone at 80 °C in the presence of some dibutyltin dilaurate as the catalyst. The polycondensation yields were not as high as expected but it was shown that the polymer was fitted with PMMA grafts, originating from the incorporation of the ω-dihydroxy-PMMA macromonomer into the polycondensate. The MMA content of the copolymer is somewhat lower than that in the feed. No data were given on molecular weights and polydispersity.

A similar procedure was used with ω-dicarboxy-PMMA macromonomers [105]. They were mixed with another diacid (sebacic acid) and reacted stoichiometrically with a diamine (p,p'-diaminodiphenylmethane and others) at 100 °C in N-methylpyrrolidone/pyridine mixtures. The formed polyamide was shown to contain PMMA grafts although the proportion of MMA was slightly lower than in the feed.

The polymers were adequately purified but no molecular weights were reported.

It can be anticipated that such step polymerizations give rise to a random distribution of the grafts.

4 Conclusions

The critical survey of the research work carried out in the field of macromonomers exemplifies the great variety of the approaches that have been chosen for the synthesis of these macromolecules fitted with polymerizable terminal groups mostly carbon-carbon double bonds.

Techniques derived from anionic or cationic "living" polymerization methods have widely been used. They are efficient because of the long lifetime of the active sites. Once polymerization is completed these sites are used for functionalization purposes. Alternately, unsaturated ionic initiators have been used but to a lesser extent because of the requirement involved that the polymerizable groups remain unscathed during the macromonomer formation. The versatile "inifer" method has also been applied to the synthesis of macromonomers.

Much attention has been focused on free-radical polymerization in the presence of transfer agents; such processes yield ω-functional precursors that can in turn be reacted with unsaturated compounds carrying an antagonist function. This is the basic principle of what was referred to as a two-step macromonomer synthesis.

It thus appears that most of the techniques developed for (or adapted to) the macromonomer synthesis were derived from methods or concepts that had been well known beforehand. There are only few exceptions the most striking of which is the self-condensation of the monoadduct between p-divinylbenzene and a diamine developed by Tsuruta and his coworkers.

A wide choice of macromonomers is now available containing acrylic, vinylic, oxyalkylenic monomer units, bearing at the chain end various polymerizable groups (most frequently styryl and methacryloyl end groups) and covering a wide range of molecular weights. The characterization of these macromonomers has usually been performed with great care, thus rendering credible the methods described.

In the future it is likely that we will observe a broadening of the possible pathways leading to macromonomer synthesis: New methods, making use of original concepts or deriving from processes which had not been developed for that purpose, will make it possible to prepare polymer chains of various chemical nature, fitted with polymerizable end groups. The scope of this new class of compounds is already quite broad and will probably widen under the specific demand for applications. This will be a challenging task.

The homopolymerization of macromonomers has not been studied extensively yet, in spite of the interest involved. The macromolecules that result are very highly branched and compact, and the segment density in the neighbourhood of the backbone chain is exceptionally high.

The chief objective of all research in the field of macromonomers is to get an easy access to a wide choice of graft copolymers. The main limitation of the earlier procedures to synthesize graft copolymers is the small number of systems to which "grafting onto" or "grafting from" techniques could be efficiently applied. The random copolymerization of a macromonomer with another monomer offers a much broader choice, and it is also much easier to carry out, in most cases by means of free-radical initiators.

In spite of the large number of papers and patents dealing with this type of reactions, it is obvious that further research in this field is still necessary. The copolymers obtained have not been subject to adequate characterization: overall composition, compositional heterogeneity are quoted in few cases only. In most patents and papers only rough indications are given of yield, composition and "apparent" molecular weight (determined by GPC with polystyrene calibration, no account being taken of the branched character of the graft copolymers). It is well known that the properties of a graft copolymer depend on the structure of the macromolecules. Macromolecules with a small number of long grafts and those with a large number of short grafts do not exhibit the same properties, and it is necessary to have precise information on the structure of the species formed for an adequate investigation of the structure-property relationships.

Finally, a few words concerning potential applications of these graft copolymers should be added. It is known that the repulsions, which generally arise between polymer species of different chemical nature, constitute the driving force towards intramolecular phase separation. The larger the molecular weight of the grafts, the higher the tendency to form two-phase systems. Each constitutent of the graft copolymer macromolecule tends to maximize contacts with itself and to minimize heterocontacts. This is why "surface accumulation" often tends to occur when a graft copolymer is mixed with a homopolymer of the same nature as the grafts (or as the backbone). The incompatible species coats the surface and thus prevents contacts between the material and the outside. This typical behaviour has led to a certain number of applications of graft copolymers as compatibilizers in polyblends, dispersing agents, adhesives, surface modifiers, coatings or varnishes, antistatic additives, wetting agents, and as materials for biomedical use.

Special interest is devoted to graft copolymers in which the backbone is hydrophobic and the grafts are hydrophilic (or vice versa). These materials are efficient water suspension stabilizers, and give rise to stable emulsions in water (and even sometimes to microemulsions in the presence of a third constituent). They can be used in many instances whenever oil-in-water or water-in-oil dispersions are needed and the number of applications is steadily increasing.

Even in polymer chemistry these species may be of interest since more and more attention is devoted to polymerization in confined volumes, such as micelles, whereby the graft copolymer could first help in the emulsion polymerization of a monomer compatible with the backbone and subsequently act as a surface modifier of the polymer formed (wetting agent, pigment binder, coating binder, antistatic compound, adhesion factor, etc.).

Judging from the increasing number of published papers it can be anticipated that research in the broad field of macromonomers will still further develop in the near future. New macromonomers will be synthesized and the existing methods of preparation can still be improved. However, the emphasis will surely be put on the easy access to a great variety of graft copolymers and on the search for new applications of these. It is our hope that adequate characterizations of the graft copolymers will be included in future research programs, for a better understanding of the structure-property relationships, and as a contribution of the very exciting macromonomer research to basic polymer science.

Acknowledgement: We wish to thank Dr. Patrick Masson for his kind help in bibliographic research and for his cooperation.

5 References

1. Cho, I.: Polym. (Korea) *4*, 210 (1980)
2. Yamashita, Y.: Kobunshi, High Polymers Japan *31*, 988 (1982); J. Appl. Polym. Sci. *36*, 193 (1981)
3. Saegusa, T.: Top. Current Chem. *100*, 75 (1982)

4. Vlegels, M.: Chem. en Kunstof 8, 12 (1982)
5. Rempp, P. et al.: Plaste und Kautschuk 28, 365 (1981)
6. Vogl, O.: J. Polym. Sci., Polym. Symp. 64, 1 (1978)
7. Greber, G., Reese, E.: Makromol. Chem. 55, 96 (1962)
8. Greber, G., Balciunas, A.: Makromol. Chem. 69, 193 (1963)
9. Greber, G., Balciunas, A.: Makromol. Chem. 79, 149 (1964)
10. Gillman, K. F., Senogles, E.: Polym. Lett. 5, 477 (1967)
11. ICI Ltd. Neth. Appl. 6, 409, 047 (1965)
12. Szwarc, M.: Carbanions, Living Polymers and Electron Transfer Processes New York, Interscience Publishers 1968
13. Waack, R., Doran, M. A.: Polym. 2, 365 (1961)
14. Waack, R.: U.S. Patent 3, 235, 626 (1966)
15. Waack, R., Doran, M. A.: J. Org. Chem. 32, 3395 (1967)
16. Masson, P. et al.: Polym. Bull. 7, 17 (1982)
17. Kobayashi, S. et. al.: Polym. Bull. 9, 169 (1983)
18. Milkovich, R.: ACS. Polym. Preprints 21, 40 (1980) Milkovich, R., Chiang, M. T.: U.S. Patent 3, 786, 116 (1974) Other Patents by Milkovich R. and Chiang M. T. include U.S. Patents 3 846 393, 3 832 423, 3 842 050, 3 842 057, 3 842 058, 3 842 059 (1974); 3 862 098, 3 862 101, 3 862 102, 3 862 267 (1975); 4 085 168 (1978)
19. Rempp, P., Loucheux, M. H.: Bull. Soc. Chim. Fr. 1958, 1497
20. Anderson, B. C. et al.: Macromol. 14, 1599 (1981)
21. Asami, R.: IUPAC 28th. Symp. On Macromolecules, Amherst 1982, Preprints p. 71
22. Freyss, D., Leng, M., Rempp, P.: Bull. Soc. Chim. Fr. 1964, 221
23. Brody, M., Richards, D. H., Szwarc, M.: Chem. and Ind. (London) 1958, 1473
24. Masson, P., Franta, E., Rempp, P.: Makromol. Chem., Rapid Commun. 3, 499 (1982)
25. Candau, F. et al.: C.R. Acad. Sci. 284, 837 (1977)
26. Sigwalt, P.: Angew. Makromol. Chem. 94, 161 (1981)
27. Nametkine, N. S., Topchiev, A. V., Dourgarian, S. G.: J. Polym. Sci. c 4, 1053 (1963)
28. a) Ph. Chaumont, Herz, J., Rempp, P.: Europ. Polym. J. 15, 537 (1979) b) Ph. Chaumont al.: Europ. Polym. J. 15, 459 (1979)
29. Kawakami, Y. et. al.: Polym. J. 14, 913 (1982)
30. Katz, D., Zewi, I. G.: J. Polym. Sci. (Chem. Ed.) 16, 597 (1978)
31. Quirk, R. P., Chien, W. C.: Makromol. Chem. 183, 2071 (1982)
32. Schulz, D. N., Sanda, J. C., Willoughby, B. G.: Anionic Polymerization, ACS Symp. Series 166, 427 (1981)
33. Graetz, C. W.: Europ. Pat. Appl. 12524 (1980)
34. Gramain, P., Frere, Y.: to be published
35. Stowe, S. C.: U.S. Pat. 3 190 925 (1965)
36. Hamaide, T. et. al.: to be published in: Nouveau Journal de Chimie
37. Spach, G., Levy, M., Szwarc, M.: J. Chem. Soc. 1962, 355
38. Königsberg, I., Jagur-Grodzinski, J.: J. Polym. Sci. (Chem. Ed.), submitted for publication
39. Dreyfuss, P., Dreyfuss, M. P.: Adv. Polym. Sci. 4, 528 (1967)
40. Penczek, S., Kubisa, P., Matyjaszewski, K.: Cationic Ring-Opening Polymerization of Heterocyclic Monomers, in: Adv. Polym. Sci. 37, 1980
41. Brown, W. B., Szwarc, M.: Trans. Farad. Soc. 54, 416 (1958)
42. Franta, E. et al.: J. Polym. Sci. Symp. 56, 139 (1976)
43. Smith, S., Hubin, A. J.: J. Macromol. Sci. (A) 7, 1399 (1973)
44. Croucher, T. G., Wetton, R. E.: Polym. 17, 205 (1976)
45. Zilliox, J. G. et al.: IUPAC Symp. Mainz 1979, Preprint p. 56
46. Sierra-Vargas, J. et al.: Polym. Bull. 3, 83 (1980)
47. Saegusa, T., Matsumoto, S.: J. Polym. Sci. (A) 6, 1559 (1968)
48. Brzezinska, K. et. al.: Makromol. Chem. 178, 2491 (1977)
49. Burgess, F. J. et. al.: Polym. 19, 334 (1978)
50. Asami, R., Takaki, M.: 27th. IUPAC Symp. on Macromolecules Strasbourg 1981, Preprints I, p. 240
51. Asami, R. et. al.: Polym. Bull. 2, 713 (1980)
52. Sierra-Vargas, J. et al.: Polym. Bull. 7, 277 (1982)

53. Abadie, M. et. al.: Polym. *23*, 445 (1982)
54. Richards, D. H.: Brit. Polym. J. *12*, 89 (1980)
55. Takaki, M., Asami, R., Kuwabara, T.: Polym. Bull. *7*, 521 (1982)
56. Reibel, L. et. al.: Polymeric Amines and Ammonium Salts (ed.) E. Goethals, p. 89, Oxford, New York, Pergamon Press 1980
57. Franta, E., Rempp, P., Reibel, L.: to be published
58. Munir, A., Goethals, E. J.: J. Polym. Sci. (Chem. Ed.) *19*, 1985 (1981)
    Goethals, E. J., Munir, A.: Polymeric Amines and Ammonium Salts (ed.) E. Goethals p. 19, Oxford, New York, Pergamon Press 1980
59. Goethals, E. J., Vlegels, M.: Polym. Bull. *4*, 521 (1981)
60. Smith, S., Schultz, W. J., Newmark, R. A.: ACS Symp. Series, *59*, 13 (1977)
61. Goethals, E. J.: 28th IUPAC Symp. on Macromolecules, Amherst 1982, Preprint p. 204
62. Tanizaki, Y. et al.: ACS/CSJ Chem. Congress, Worldwide Prog. of the Petrochem. Org. and Polym. Chem. Indust. Honolulu, 1979
63. Heitz, W., Kress, H. J., Stix, W.: 28th. IUPAC Symp. on Macromolecules. Amherst 1982, Preprint p. 74
64. Kress, H. J., Heitz, W.: Makromol. Chem.. Rapid Commun. *2*, 427 (1981)
65. Kennedy, J. P.: Cationic Polymerization of Olefins: A Critical Inventory, New York, Wiley Intersci. Publ. 1974
    Kennedy, J. P., Marechal, E.: Carbocationic Polymerization, New York, Wiley Intersci. Publ. 1982
66. Kennedy, J. P.: Polym. J. *12*, 609 (1980)
67. Kennedy, J. P.: 5th. Int. Symp. on Cationic Polym., Kyoto 1980, Preprints p. 6
68. Kennedy, J. P., Smith, R. A.: ACS Polym. Prepr. *20*, 316 (1979)
69. Kennedy, J. P., Frisch, K. Jr.: IUPAC Symp. on Macromolecules, Firenze 1980 Preprints II, 162
70. Chang, V., Kennedy, J. P.: Polym. Bull. *5*, 379 (1981)
71. Ivan, B., Kennedy, J. P., Chang, V.: J. Polym. Sci. (Chem. Ed.) *18*, 1523, 1539, 3177 (1980)
72. Kennedy, J. P., Liao, J. P.: Polym. Bull *5*, 11 (1981); *6*, 135 (1981)
73. Percec, V. et al.: Polym. Bull. *8*, 25 (1982)
74. Kennedy, J. P. et al.: Polym. Bull. *1*, 575 (1979)
75. Tessier, M.: Thesis, Paris 1982
76. Kennedy, J. P., Huang, S. Y., Smith, R. A.: Polym. Bull. *1*, 371 (1979); J. Macromol. Sci. (Chem.) *(A)14*, 1085 (1980)
77. Kennedy, J. P., Lo, C. Y.: ACS Polym. Prepr. *23*, 99 (1982)
78. Kennedy, J. P. et al.: Polym. Bull. *8*, 551, 557, 563, 571 (1982)
79. Narita, T., Yamagushi, Y., Tsuruta, T.: Bull. Chem. Soc. Japan *46*, 3825 (1973)
80. Imai, N., Narita, T., Tsuruta, T.: Tetrahedron Lett. *38*, 3517 (1971); Bull. Chem. Soc. Japan *46*, 1242 (1973)
81. Tsuruta, T.: Polymeric Amines and Ammonium Salts (ed.) Goethals, E. J., p. 163, Oxford, New York, Pergamon Press 1980
82. Tsuruta, T., Narita, T., Nitadori, Y.: Makromol. Chem. *177*, 3255 (1976)
83. Nitadori, Y., Tsuruta, T.: Makromol. Chem. *180*, 1877 (1979)
84. Nishimura, T. et. al.: Makromol. Chem., Rapid Commun. *1*, 573 (1980); Makromol. Chem. *183*, 29 (1982)
85. Tsuruta, T.: 28th IUPAC Symp. on Macromolecules, Amherst 1982, Preprint p. 73
86. Maeda, M., Nitadori, Y., Tsuruta, T.: Makromol. Chem. *181*, 2245 (1980)
87. Maeda, M., Nitadori, Y., Tsuruta, T.: Makromol. Chem. *181*, 2251 (1980)
88. Maeda, M., Inoue, S.: Makromol. Chem., Rapid Commun. *2*, 537 (1981)
89. Kimura, M. et. al.: Makromol. Chem. *183*, 1393 (1982)
90. Ferruti, P. et. al.: J. Polym. Sci. (Chem. Ed.) *15*, 2151 (1977)
91. Osmond, D. W., Waite, F. A., Walbridge, D. J.: Brit. Pat. 1 122 397 (1968)
92. Hudecek, S. et. al.: Polym. Bull. *3*, 143 (1981)
93. Thompson, M. W., Waite, F. A.: U.S. Pat. 3 390 206 (1968) U.S. Pat. 3 627 239 (1971); Brit. Pat. 10 96 912 (1967)
94. Konter, W. et. al.: Makromol. Chem. *182*, 2619 (1981)
95. Bamford, C. H., Jenkins, A. D.: Nature *176*, 78 (1955)

96. Bamford, C. H., Jenkins, A. D., Johnston, R.: Trans. Farad. Soc. *55*, 179 (1959)
97. Brosse, J. C., Baunier, M., Legeay, G.: Makromol. Chem. *182*, 3457 (1981); *183*, 303 (1982)
99. Yamashita, Y. et. al.: Polym. Bull. *5*, 335 (1981)
98. Ito, K., Usami, N., Yamashita, Y.: Macromol. *13*, 216 (1980)
100. Waite F. A.: J. Oil Col. Chem. Ass. *54*, 342 (1971)
101. Yamashita, Y. et al.: Polym. J. *14*, 255 (1982)
102. Kawakami, Y. et al.: J. Polym. Sci. (Polym. Lett.) *19*, 629 (1981)
103. Simms, J. A., Walus, A. N.: U.S. Pat. 3 716 506 (1973)
104. Ashby, T. L., Parrish, D. B.: U.S. Pat. 3 480 601 (1969)
105. Yamashita, Y. et. al.: Polym. Bull. *5*, 361 (1981)
106. Chujo, Y., Tastuda, T., Yamashita, Y.: Polym. Bull. *8*, 239 (1982)
107. Asher, M., Vofsi, D.: J. Chem. Soc. *(B) 1963*, 1887 *1964*, 4962
108. Boutevin, B., Pietrasanta, Y.: J. Polym. Sci. (Chem. Ed.) *19*, 499, 511 (1981)
109. Boutevin, B. et. al.: Europ. Polym. J. *14*, 353 (1978)
110. Boutevin, B., Pietrasanta, Y., Taha, M.: Makromol. Chem. *183*, 2977, 2985, 2995 (1982)
111. El Sarraf, T. et. al.: to be published in: Die Makromolekulare Chemie
112. Boutevin, B., Pietrasanta, Y., Sideris, A.: J. Fluorine Chem. *20*, 727 (1982)
113. Kahrs, K. H., Zimmermann, J. W.: Makromol. Chem. *58*, 75 (1962), see also Bui Duc Hao, Thesis, Strasbourg 1978
114. Grubisic, Z., Rempp, P., Benoit, H.: J. Polym. Sci., Polym. Let. *5*, 753 (1967)
115. Asami, R. et. al.: Polym. J. *13*, 141 (1981)
116. Zilliox, J. G.: Makromol. Chem. *156*, 121 (1972)
117. Sierra-Vargas, J., Franta, E., Rempp, P.: Makromol. Chem. *182*, 2603 (1981)
118. Kennedy, J. P., Lo, C. Y.: Polym. Bull. *8*, 63 (1982)
119. Revillon, A., Hamaide, T.: Polym. Bull. *6*, 235 (1982)
120. Kelen, T., Tüdös, F.: J. Macromol. Sci. Chem. *9*, 1 (1975)
121. Rempp. P., Franta, E.: J. Herz, ACS Symp. Series, Anionic Polymerization (ed.) McGrath, p. 59, 1981
122. Gallot, Y., Rempp, P. Parrod. J.: J. Polym. Sci., Polym. Lett. *1*, 329 (1963)
123. Candau, F., Afshar-Taromi, F., Rempp, P.: Polym. *18*, 1253 (1977)
124. Rempp, P., Franta, E.: Pure Appl. Chem. *30*, 229 (1972)
125. Kennedy, J. P.: Cationic Graft Copolymerization, Appl. Polym. Symp., J. Appl. Polym. Sci. *30*, (1977)
126. Smets, G., Beylen, M. van: Makromol. Chem. *69*, 140 (1963)
127. Bamford C. H. et al.: Polym. *10*, 759, 771, 885 (1969); *13*, 57 (1972)
128. Masson, P., Rempp, P., Franta, E.: to be published
129. Masson, P.: Thesis, Strasbourg 1982
130. Asami, R. et. al.: Polym. J. *13*, 141 (1981)
131. Schulz, G. O., Milkovich, R.: J. Appl. Polym. Sci. *27*, 4773 (1982)
132. Yamashita, Y.: 28th. IUPAC Symp. on Macromolecules, Amherst. 1982, Preprint p. 93
133. Owen, M. J., Kendrick, T. C.: Macromol. *3*, 458 (1970)
134. Braun, J. M., Guillet, J. E.: Adv. Polym. Sci. *21*, 107 (1976); Gray, D. G.: Progr. Polym. Sci. *5*, 1 (1977)
135. Nishimura, T. et. al.: Makromol. Chem., Rapid Commun. *1*, 573 (1980)
136. Mathias, L. J., Canterberry, J. B., South, M.: J. Polym. Sci., Polym. Lett. *20*, 473 (1982)

T. Saegura (Editor)
Received June 8, 1983 ...

Linear Amino Polymers:
Synthesis, Protonation and Complex Formation

Paolo Ferruti
Istituto di Chimica degli Intermedi dell' Università,
Viale Risorgimento 4, 40136 Bologna, Italy

Rolando Barbucci
Istituto di Chimica Generale dell' Università,
Piano dei Mantellini 44, 53100 Siena, Italy

Linear amino polymers containing basic nitrogen atoms are critically reviewed with regard to their synthesis, protonation and complex formation in solution with metal ions. Cross linked resins having essentially the same structure as linear polymers, are also mentioned. As far as the protonation is concerned, special care has been given to thermodynamic aspects, and to the most probable protonation mechanism. Complexing abilities of these polymers have been evaluated through stability constants and spectroscopic parameters. Practical implications of the properties have been considered.

List of Abbreviations and Symbols

$$pK \qquad = -\log \frac{[H^+][L]}{[HL^+]}$$

pK_a = pK referred to coordinated acid dissociation

pK_0 = pK of ionizing groups in the absence of electrostatic interactions with other ionizing groups

$$pK_{app} \qquad = pH + \log \frac{\alpha}{1-\alpha}$$

pK_{av} = average pK

pH_p = pH at beginning polymer precipitation

α = degree of ionization; degree of neutralization

$\log \beta'$ = logarithm of the overall stability constant of complexes

γ = fraction of ionizable groups present in the form of charged neighbour pairs

β = degree of protonation of residual tertiary amino groups

δ = overall degree of neutralization

ε = degree of neutralization of quaternary ammonium hydroxide sites

(β, δ, ε are taken from references 69 and 70)

M   = Mol, molar

c   = concentration; cM = molar concentration; cM_{ppt} = molar concentration at the precipitation point

$\bar{M}n$ = number average molecular weight

$\langle P \rangle$ = average polymerization degree

$[\eta]$ = viscosity number; intrinsic viscosity

η_{sp} = specific viscosity

η_{inh} = inherent viscosity at a given concentration

ΔH^0, ΔG^0, ΔS^0 = enthalpy, free energy, and entropy of formation, respectively

DMF   = dimethyl formamide

ORD   = optical rotatory dispersion

CD   = circular dichroism

Please note: some of the symbols and abbreviations have been modified with respect to the original papers in order to avoid superposition of significance.

1 Introduction

The aim of this review is to relate on the most recent developments concerning the chemistry of polymeric amines. Only those families of polymeric amines which have been extensively studied, and on which a sufficient amount of data is available to allow a fairly complete picture of their behaviour in solution, will be considered.

Therefore, the following families of polymers have been chosen:

Poly(vinylamine) (PVA)

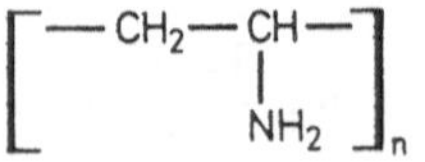

$$\left[\!-CH_2-CH-\!\right]_n \quad \overset{|}{NH_2}$$

Poly(iminoethylene) or poly(ethylene imine) (PEI)*

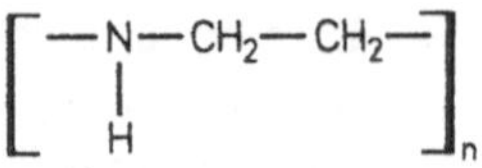

$$\left[\!-\underset{H}{\overset{|}{N}}-CH_2-CH_2-\!\right]_n$$

Poly(N-methyl-ethylene-imine) (PMEI) and related polymers

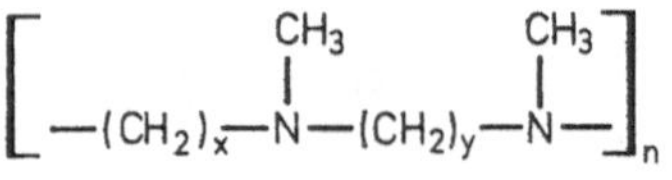

$$\left[\!-(CH_2)_x-\underset{}{\overset{CH_3}{\overset{|}{N}}}-(CH_2)_y-\underset{}{\overset{CH_3}{\overset{|}{N}}}-\!\right]_n$$

Poly[thio-1-(N,N-dialkyl-aminomethyl)-ethylene] (PTN)

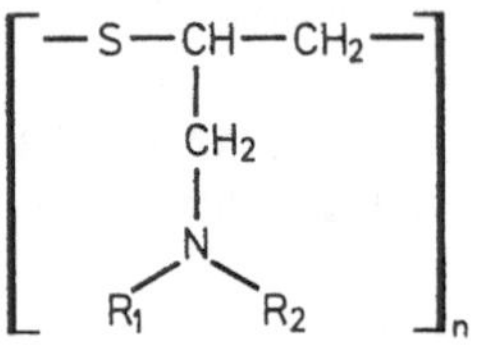

$$\left[\!-S-CH-CH_2-\!\right]_n$$
$$\overset{|}{CH_2}$$
$$\underset{R_1 \quad R_2}{\overset{|}{N}}$$

Poly(4- or 2-vinylpyridine) (P4VP) or (P2VP)

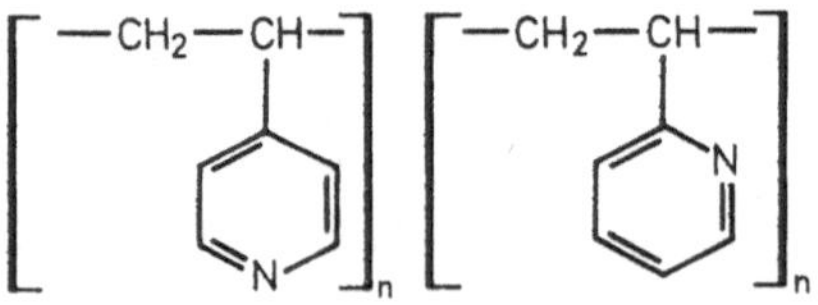

$$\left[\!-CH_2-CH-\!\right]_n \quad \left[\!-CH_2-CH-\!\right]_n$$

Poly(N-vinyl-imidazole) (PVI), and poly[4(5)-vinyl-imidazole] (P4VI)

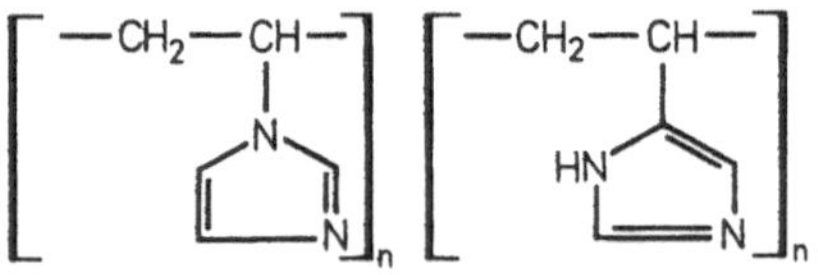

$$\left[\!-CH_2-CH-\!\right]_n \quad \left[\!-CH_2-CH-\!\right]_n$$

Poly(amido-amine)s (PAA)

$$\left[\!-CH_2-CH_2-\overset{O}{\overset{||}{C}}\underset{R_1}{\overset{|}{N}}-R_2-\underset{R_3}{\overset{|}{N}}\overset{O}{\overset{||}{C}}-CH_2-CH_2-\underset{R_4}{\overset{|}{N}}-R_5-\underset{R_6}{\overset{|}{N}}-\!\right]$$

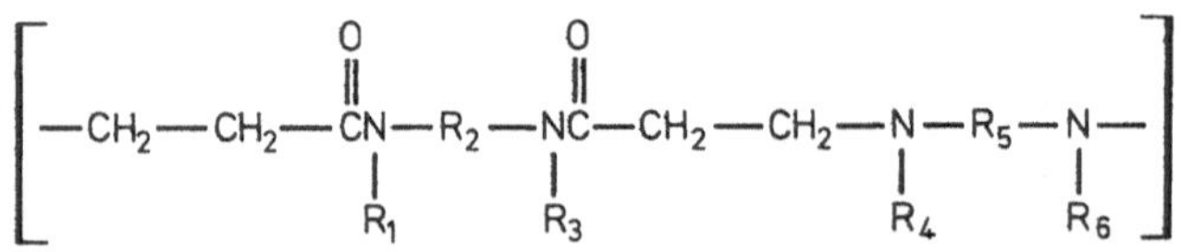

Poly(diaza-crown ether)s (PMC) (MC=macrocyclic ligand)

$$\sim\!\!\!\text{—(MC)—R—(MC)—R—(MC)—R—}\!\!\!\sim$$

We do not intend to consider basic poly(α-amino acid)s, such as polylysine or polyhistidine, among the main subjects of this review, since they have been the object of many other reviews, along with other poly(α-amino acid)s.

In addition, some cross-linked resins having essentially the same structure as linear polymers belonging to the above list will be considered. All available data relative to each class of polymers will be reported, and critically evaluated.

---

* In the literature, linear polymers are referred to as poly(iminoethylene), and branched polymers as poly(ethylenimine), since the latter derive from the ring-opening polymerization of ethylenimine. In this review, both polymers will be referred to as PEI. Mention will be made, however, to their structure whenever opportune.

2 Synthesis

2.1 Poly(vinylamine) (PVA)

As a matter of course, *polyvinyl-amine* cannot be prepared from the corresponding monomer. Therefore, this polymer must be prepared from suitable polymeric precursors. The first way involves the hydrazinolysis of poly(N-vinyl-phthalimide) [1].

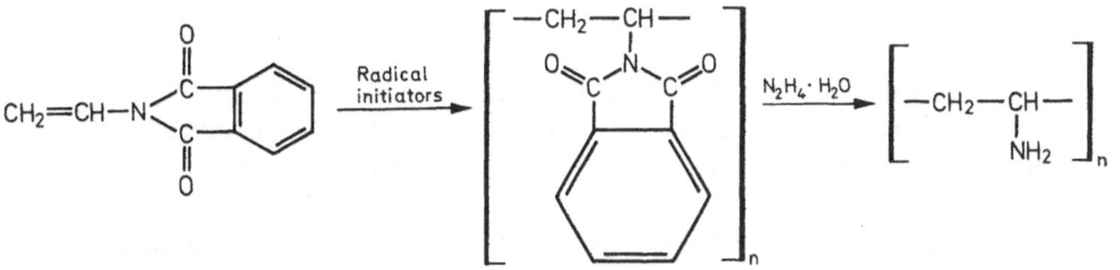

A Schmidt reaction performed on poly(acrylic acid) has been also proposed [2]. Better methods involve the hydrazinolysis of poly(N-vinyl-succinimide) [3], the hydrolysis of poly(N-vinyl-tert-butylcarbamate) [4]:

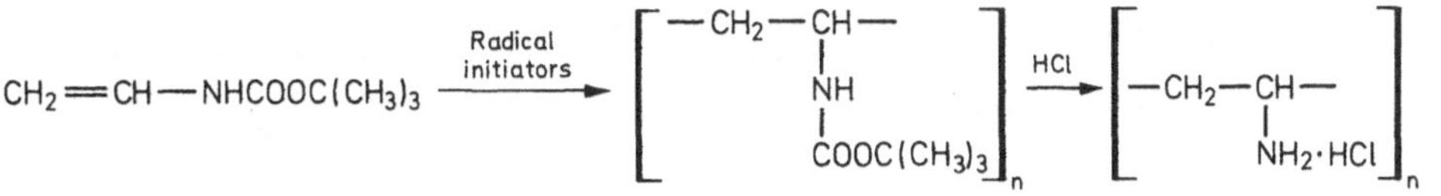

and the hydrolysis of poly(N-vinyl-acetamide) [5]

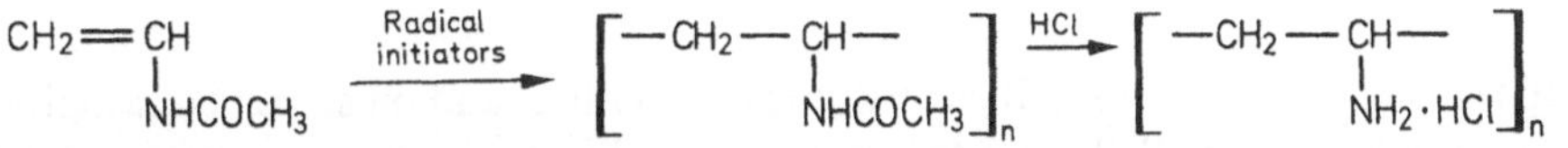

A conversion degree of more than 95% was achieved by the poly(N-vinyl succinimide) and poly(N-vinyl acetamide) methods, while by the poly(N-vinyl carbamate) method essentially pure poly(vinylamine) was obtained.

Cross-linked poly(vinyl-amine) was obtained by hydrazinolysis of a cross-linked poly(N-vinylphthalimide), in turn obtained by radical copolymerization of N-vinyl-phthalimide and divinylbenzene [6].

2.2 Poly(ethylene imine) (PEI)

Branched *poly(ethyleneimine)* is usually obtained by ring opening polymerization of ethyleneimine with cationic initiators. The various aspects of this reaction have been reviewed elsewhere [7]. This is the origin of commercial PEI. For this kind of PEI, it has been definitely established that the molecular structure is highly branched, independently of the modalities of preparation. As a result, there are three types of functional groups, i.e. primary, secondary and tertiary amino groups. For instance, studies [8] on commercial PEI's proved that of the total amount of functional amino groups about 25% is primary, 50% secondary and 25% tertiary. As a consequence, conventional PEI is amorphous, because many branches along the polymer chain inhibit crystallization.

Linear PEI can be obtained by cationic ring-opening polymerization of 2-oxazoline, followed by alkaline hydrolysis of the resulting poly(N-formylethyleneimine) [9].

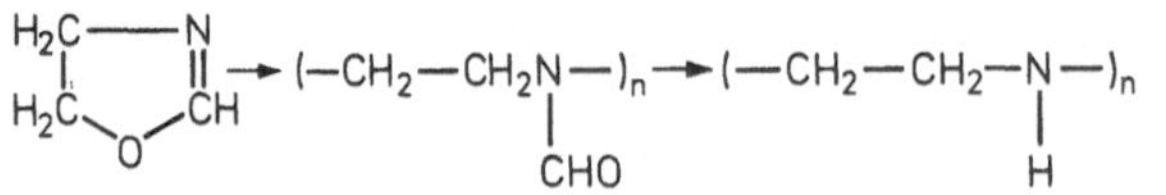

2-Alkyl substituted 2-oxazolines, for instance 2-methyl-2-oxazoline [9], or 2-ethyl-2-oxazoline [10] can be used in the place of unsubstituted 2-oxazoline. In all cases, the molecular weight of the resulting PEI is relatively low, usually in the order of a few thousands.

The cationic ring-opening bulk polymerization of 2-phenyl-2-oxazoline at 133 °C, using dimethyl sulphate as catalyst, gave a poly(N-benzoylethyleneimine), with a considerably high molecular weight (up to 4×10^5 dalton) [11]. Poly(N-benzoylethylene-imine) is very resistent to alkaline hydrolysis, but can be hydrolized under acidic conditions giving linear high molecular weight PEI. [12]

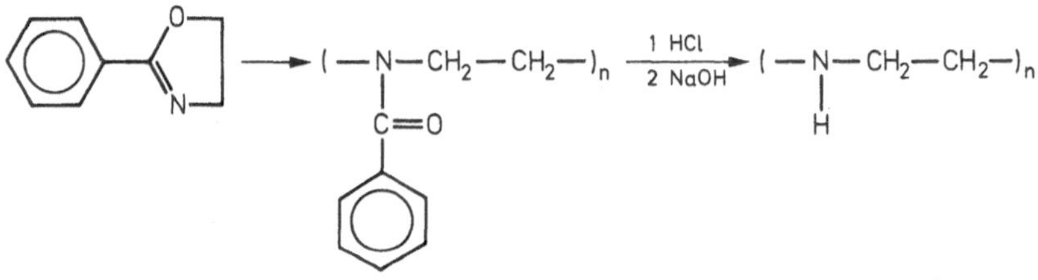

Owing to its regular structure, linear PEI may crystallize, and on stretched samples of high molecular weight PEI an x-ray crystallographic study could be performed. It seems to crystallize as arrays of double strand helices [12]. Furthermore, two crystalline hydrates of linear PEI of $\bar{M}n$ about 2000 have been reported [13].

Crosslinked resins of polyethylene imine structure have been prepared by various methods, including the reactions of polyethylene imine with epichlorohydrine [14], with allyl chloride [15], with ethylene dibromide [16] or dichloride [17], and with toluene diisocyanate [18].

2.3 Poly(N-methyl-ethylene imine) (PMEI) and related Polymers

Poly(N-alkyl-ethylene imine) can be prepared by cationic ring-opening polymerization of N-alkylaziridines; however, branched polymers are obtained by this way [19]. Linear poly-MEI may be obtained by N-methylation of linear PEI by the Clarke-Eschiweiler variation of the Leukart reaction, in which an excess of formic acid is used together with formaldehyde. The reaction may be performed either on PEI, or directly on poly(N-formyl ethylene imine) [20], which is the product of the polymerization of 2-oxazoline (see above).

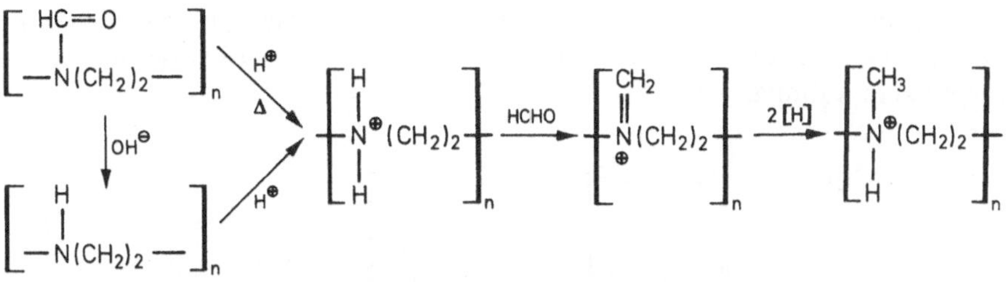

Higher homologs of PEI can be obtained by cationic ring-opening polymerization of azetidine monomers [21]. Polymers of this kind can also be N-methylated as previously described in the case of PEI. The cationic polymerization of N-substituted azetidines may give rise to poly(tertiary amine)s. Linear poly(tertiary amine)s have been obtained by selective dealkylation of poly(quaternary ammonium salt)s [22].

2.4 Poly[thio-1-(N,N-dialkyl-aminomethyl)-ethylene]s (PTN)

Poly[thio-1-(N,N-alkyl-aminomethyl)-ethylenes] can be obtained by polymerization of N,N-dialkyl[N(thiirane-2-methyl)]amines with $ZnEt_2$—CH_3OH as initiator system [23,24].

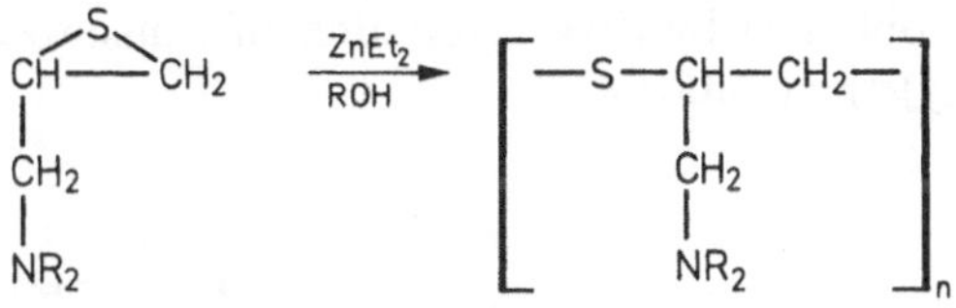

This type of monomer gives two possibilites of obtaining optically active polymers. Owing to the presence of a chiral center in the main chain, it is possible to obtain optically active polymers either by polymerizing a racemic monomer by a stereoselective initiator [25], or by polymerizing enantiomeric monomers [26,27]. On the other hand, it is also possible to obtain optically active polymers by polymerizing monomers having an asymmetric carbon on a nitrogen substituent, as in the case of N-sec-butyl-N-methyl-N(thiiranyl-2-methyl)amine [24,28,29].

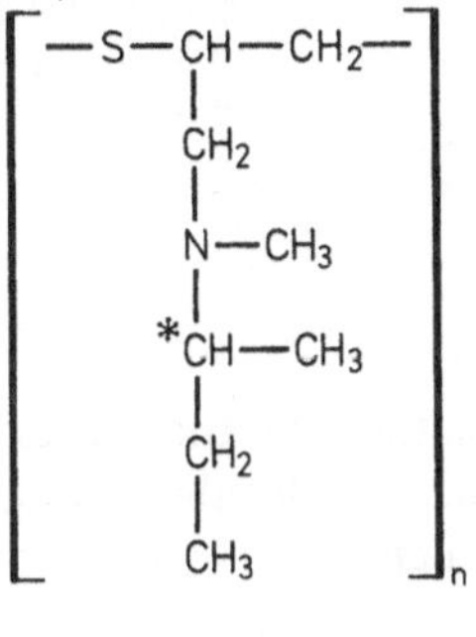

2.5 Poly(vinyl-pyridine)s (P4VP, P2VP)

The preparation of poly(2-vinyl-pyridine), by radical or anionic polymerization of the corresponding monomer, has been described long time ago. Atactic or isotactic, partially crystallizable polymers may be obtained [30]. Poly(4-vinyl-pyridine) may be prepared in a similar way [31].

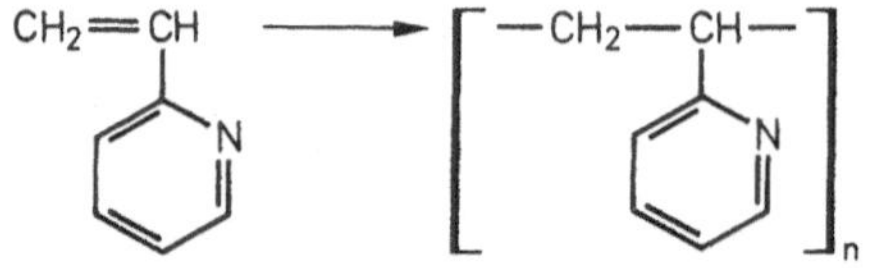

Cross-linked resins of 2- or 4-vinyl-pyridine structure have been prepared by copolymerizing 2- or 4-vinyl-pyridine with divinylbenzene [32].

Poly(4-vinyl pyridine) was also cross-linked with α, ω dibromo alkanes, such as for instance 1,4 dibromobutane. In this case, the crosslinking reaction may be performed in the presence of metal ions, thus obtaining resins showing a selectivity in the absorption of the ions which were present during their preparation [33].

2.6 Poly(N-vinyl-imidazole) (PVI) and Poly[4(5)-vinyl-imidazole] (P4VI)

Poly[4(5)vinyl-imidazole] can be obtained by radical polymerization of the corresponding monomer, in turn obtained by decarboxylation of transurocanic acid [34] or by pyrolysis of 4(β-acetoxy)ethylimidazole [35].

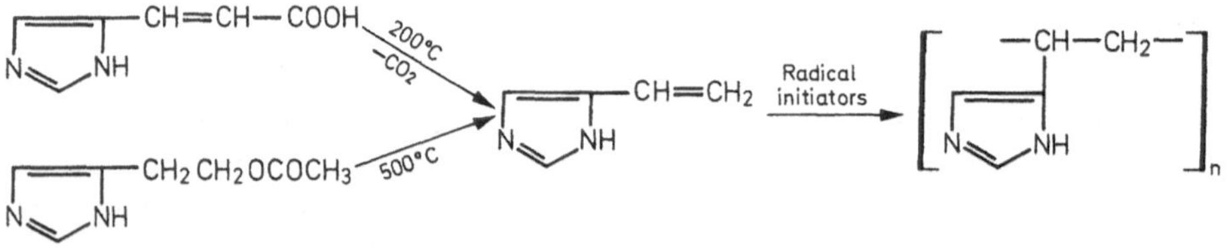

Poly N(vinyl-imidazole) can be obtained by radical polymerization of the corresponding monomer [36], which is a commercial product.

Cross-linked poly(N-vinyl-imidazole) was obtained by reaction of poly(N-vinyl-imidazole) with 1,4 dibromobutane [37].

2.7 Poly(amino-amine)s (PAA)

Poly(amido-amine)s are a class of polymers characterized by the presence of amido and amino groups regularly arranged along the macromolecular chain. Additional functional groups may be introduced as side substituents.

Linear poly(amido-amine)s are obtained by polyaddition of primary monoamines, or bis-(secondary amines), to bis-acrylamides [38-44].

a) $n\ CH_2=CH-CO-\underset{\underset{R^1}{|}}{N}-R^2-\underset{\underset{R^3}{|}}{N}-CO-CH=CH_2 + n\ H_2N-R^4 \rightarrow$

$$\rightarrow \left[CH_2-CH_2-CO-\underset{\underset{R^1}{|}}{N}-R^2-\underset{\underset{R^3}{|}}{N}-CO-CH_2-CH_2-\underset{\underset{R^4}{|}}{N} \right]_n$$

b) $n\ CH_2=CH-CO-\underset{\underset{R^1}{|}}{N}-R^2-\underset{\underset{R^3}{|}}{N}-CO-CH=CH_2 + n\ H\underset{\underset{R^4}{|}}{N}-R^5-\underset{\underset{R^6}{|}}{N}H \rightarrow$

$$\rightarrow \left[CH_2-CH_2CO-\underset{\underset{R^1}{|}}{N}-R^2-\underset{\underset{R^3}{|}}{N}-CO-CH_2-CH_2-\underset{\underset{R^4}{|}}{N}-R^5-\underset{\underset{R^6}{|}}{N} \right]_n$$

The reaction takes place readily in water or alcohols, at room temperature, and without added catalysts. Aprotic solvents are not recommended, if high molecular weight products have to be obtained. The above method is a general one, as far as aliphatic or cycloaliphatic amines are concerned. Under the same conditions, aromatic amines do not give high polymers. Many poly(amido-amine)s are highly crystalline. Some crystallinity is even observed in some cases, where a certain degree of irregularity has been introduced.

Poly(amido amine)s carrying additional functions as side substituents can be easily obtained, starting with the appropriate monomers [38,44]. In fact, hydroxyl groups, tertiary amino groups, allyl groups, etc., if present in the monomers, do not interfere with the polymerization process.

It may be added that polymers of related structures can be obtained by substituting either bis-(acrylic esters), or divinylsulfone, for bis-acrylamides [45], or hydrazines [46] for amines. Some of these polymers are listed in Table 1.

Cross-linked resins of polyamidoamine structure can be prepared in the same way as the linear polymers by partly substituting α, ω-diaminoalkanes for the same quantity of difunctional aminic monomers in the polymerization process. According to the general scheme of poly(amido-amine)synthesis, bis(primary amine)s, having four mobile hydrogens, act as tetrafunctional monomers [47].

2.8 Poly(diaza-crown ether)s (PMC)

The synthesis of polymers containing diaza-crown ether groups in the main chain, i.e., poly(diaza-crown ether)s, has been recently achieved by incorporating these groups into macromolecular structures of conventional type, such as epoxy polymers [48,49,50], polyethers, and polyamides [51]. In the former case, water soluble polymers could be obtained.

Resins containing poly(macrocyclic)chains attached to a crosslinked polystyrene matrix have been prepared, and studied as anion exchangers in liquid chromato-

Table 1. Some Examples of Poly(amido-Amines) and Related Polymers

Structure of the repeating unit	$[\eta]$ (a) (dl/g)
$-(CH_2)_2CON\!\!\diagdown\!\!\diagup NCO(CH_2)_2N-$ $\quad\quad\quad\quad\quad\quad\quad\quad\quad\;\; CH_3$	0.40
$-(CH_2)_2CON(CH_2)_2NCO(CH_2)_2N-$ $\quad\quad\quad\quad\;\; C_2H_5 \quad C_2H_5 \quad\quad CH_3$	0.20
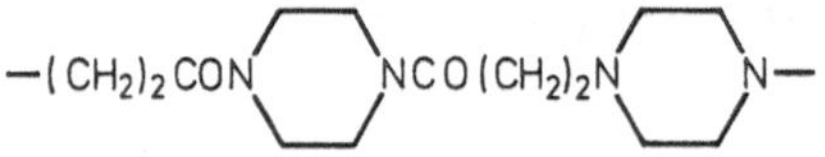 $-(CH_2)_2CON\!\!\diagdown\!\!\diagup NCO(CH_2)_2N\!\!\diagdown\!\!\diagup N-$	0.46 [b]
$-(CH_2)_2CON\!\!\diagdown\!\!\diagup NCO(CH_2)_2N(CH_2)_2N-$ $\quad\quad\quad\quad\quad\quad\quad\quad\quad\quad CH_3 \quad\quad CH_3$	0.41
$-(CH_2)_2CON\!\!\diagdown\!\!\diagup NCO(CH_2)_2N(CH_2)_2N(CH_2)_2N-$ $\quad\quad\quad\quad\quad\quad\quad\quad\quad\quad CH_3 \quad\; CH_3 \quad\;\; CH_3$	0.27
$-(CH_2)_2CON\!\!\diagdown\!\!\diagup NCO(CH_2)_2N(CH_2)_3N-$ $\quad\quad\quad\quad\quad\quad\quad\quad\quad\quad CH_3 \quad\quad CH_3$	0.43
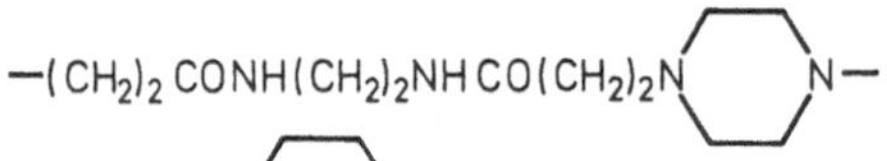 $-(CH_2)_2CONH(CH_2)_2NHCO(CH_2)_2N\!\!\diagdown\!\!\diagup N-$	0.12 [c]
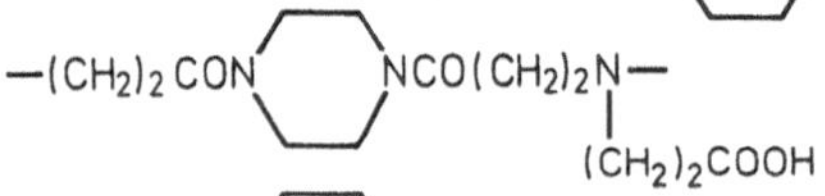 $-(CH_2)_2CON\!\!\diagdown\!\!\diagup NCO(CH_2)_2N-$ $\quad\quad\quad\quad\quad\quad\quad\quad\quad (CH_2)_2COOH$	0.27 [d]
$-(CH_2)_2CON\!\!\diagdown\!\!\diagup NCO(CH_2)_2N-$ $\quad\quad\quad\quad\quad\quad\quad\quad\quad CH_2CH=CH_2$	0.33
$-(CH_2)_2CON\!\!\diagdown\!\!\diagup NCO(CH_2)_2N-$ $\quad\quad\quad\quad\quad\quad\quad\quad\quad (CH_2)_2N(CH_3)_2$	0.19
$-(CH_2)_2CON\!\!\diagdown\!\!\diagup NCO(CH_2)_2N-(CH_2)_2-N-$ $\quad\quad\quad\quad\quad\quad\quad (CH_2)_2 \quad\quad\quad (CH_2)_2$ $\quad\quad\quad\quad\quad\quad\quad N(CH_3)_2 \quad\quad N(CH_3)_2$	0.25
$-(CH_2)_2CON\!\!\diagdown\!\!\diagup NCO(CH_2)_2N-(CH_2)_2-N-$ $\quad\quad\quad\quad\quad\quad CH_2 \quad\quad\quad\quad CH_2$ (pyridyl) (pyridyl)	0.18
$-(CH_2)_2COO(CH_2)_2OOC(CH_2)_2N\!\!\diagdown\!\!\diagup N-$	0.41

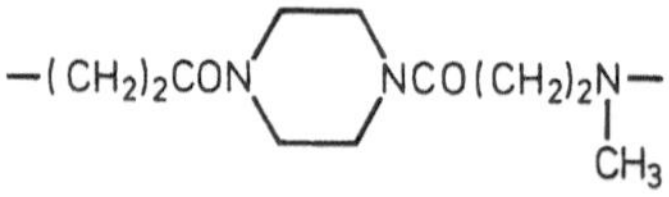

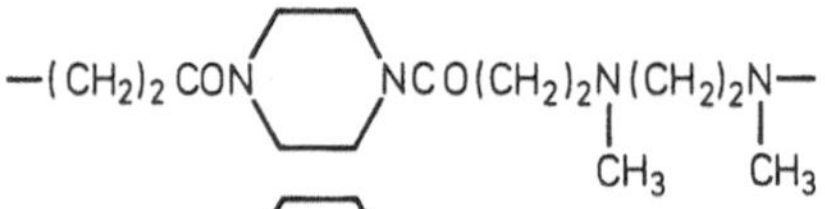

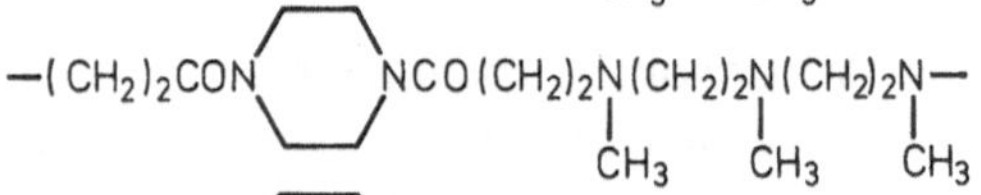

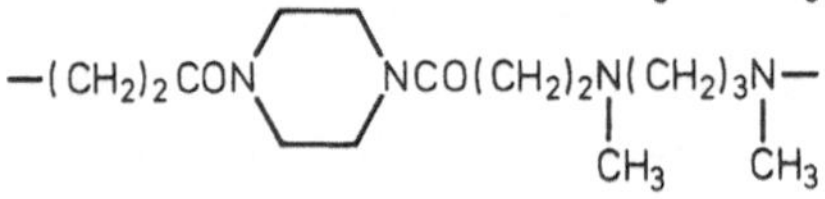

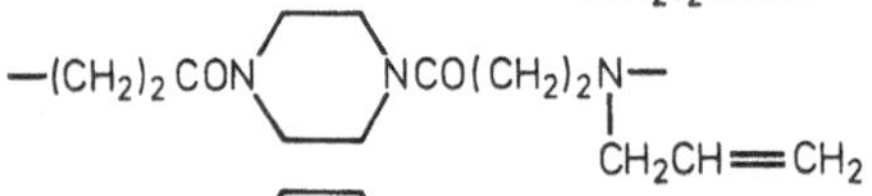

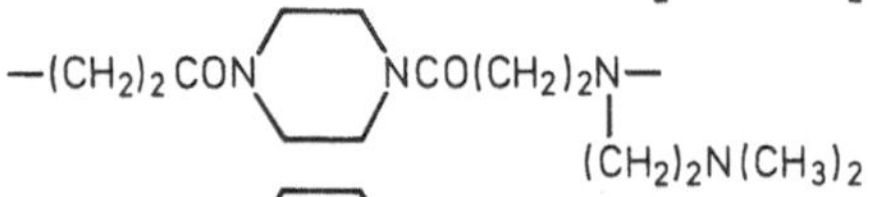

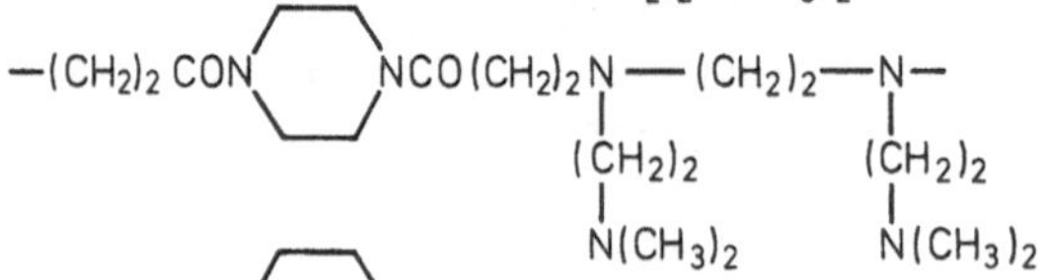

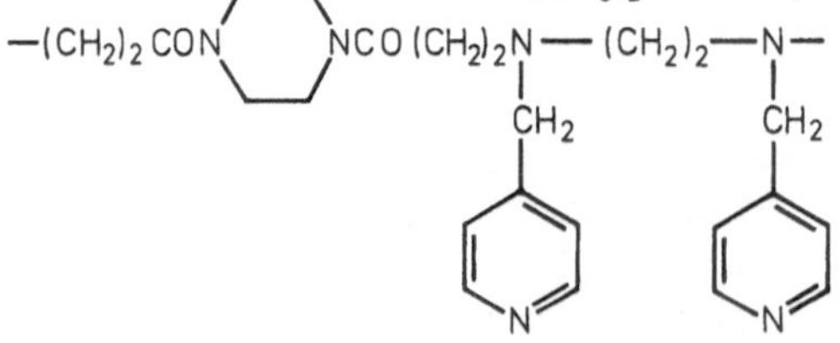

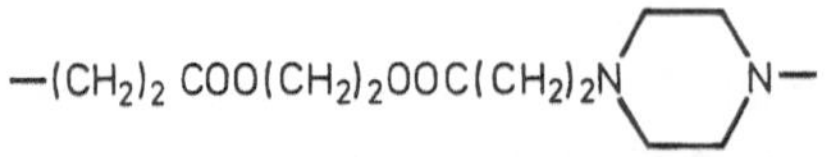

Table 1 (continued)

Structure of the repeating unit	$[\eta]^{(a)}$ (dl/g)
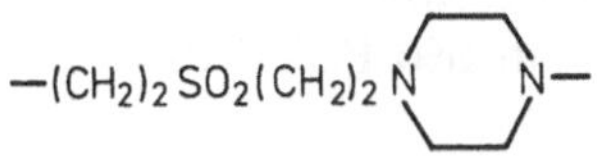 $-(CH_2)_2 SO_2 (CH_2)_2\ N\overbrace{}N-$	$0.60^{(e)}$
$-(CH_2)_2 SO_2 (CH_2)_2 \underset{CH_3}{N} (CH_2)_2 - \underset{CH_3}{N} -$	0.19
$-(CH_2)_2 \underset{CH(CH_3)_2}{CON} (CH_2)_2 - \underset{CH(CH_3)_2}{N} - CO(CH_2)_2 NH NH -$	0.16
$-(CH_2)_2 CON\overbrace{}N CO(CH_2)_2 \underset{N(CH_3)_2}{N} -$	0.10

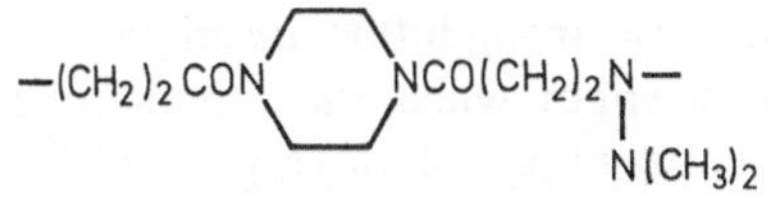

[a] in chloroform at 30°; [b] in aq. 0.1 M HCl/1 M NaCl; [c] in aq. 0.1 M CH$_3$COOH/1 M CH$_3$COONa; [d] η inh (c = 0.5 g/l) in 90% methanol/10% water at 30°; [e] in dimethylsulfoxide at 100°

graphy [52]. These ion exchange resins based on macrocyclic groups have been prepared by grafting in DMF at 60°, and in the presence of NaHCO$_3$, the monomeric or polymeric ligands on a Merrifield polymer containing 12.4% of Cl, according to the classical method of alkylation of the OH groups born by the ligands.

It has to be noted that cross-linked resins containing the functional groups which are characteristic of all the polymeric compounds reported in this Section may be prepared in a similar way, by functionalizing previously prepared crosslinked resins, usually of polystyrene type [53].

3 Protonation and Complex Formation

3.1 PVA, PEI and PMEI

The potentiometric titration curves of several poly(electrolyte)s, among which PVA and PEI (branched), have been extensively studied by Bloys von Treslong [54]. He assumes that in the protonation process, the interactions between the various aminic groups present in the macromolecule may result in a charge distribution which, at partial neutralization, is not random.

In the case of a random distribution of charges, the potentiometric curves can be interpreted on the basis of this equation:

$$pH = pK_0 + \log \frac{1 - \alpha}{\alpha} - \Delta pK$$

where pK represents the negative logarithm of the ionization constant, pK_0 the intrinsic ionization constant characteristic of the single ionizable group of the polymer, and α the ionization degree; therefore the deviations from a potentiometric curve of a monoprotic compound is accounted for in the term ΔpK.

The interpretation of the titration curves of PVA and branched PEI leeds to the following equation:

$$pK = pK_0 - \log \left(\frac{1-\alpha}{\alpha}\right)^2 \left(\frac{\alpha - \gamma}{1 - 2\alpha + \gamma}\right) + \Delta pK$$

where the second term is called the nearest neighbour interaction form (NNI). [54]

In the case of PVA a strong interaction occurs between neighbouring charged sites and, consequently, the ionic strength of the medium has a strong influence on pK_0 [55] (Table 2). Less influence is exerted by the ionic strength when the interactions between neighbouring units are smaller, as, e.g., in the case of poly(4-vinylpyridine) [56].

Table 2. pK_0 values of PVA compared with the pK value of its corresponding non-macromolecular model

Polymer	Salt conc (M)	pK_0	corresponding models	pK
PVA	0	8.6	$(CH_3)_2CHNH_2$	10.72
	1	9.5		

Branched PEI has a very compact structure in solution, as evidenced by its reduced viscosity at low ionic strength which only slightly increases by increasing the charge density of the polymer [54]. Moreover, the value of a in the intrinsic viscosity/molecular weight equation $[\eta] = K\bar{M}_n^a$ is very low (0.39) [57] as compared with that of most linear polymers (0.6–0.8).

The interactions between the various units of branched PEI, both charged and uncharged, are strong enough to make the interpretation of the titration curves of this polymer rather questionable [54], even assuming that the primary amino groups are protonated first, and that a tautomeric equilibrium occurs among the different basic nitrogens [54].

3.1.1 Protonation

Calorimetric data on the protonation of PVA, and linear PEI at various temperatures and ionic strengths have been published by St. Pierre et al. [10]. The very important point of this study is, that ΔH^0 data allow a better evaluation of the polyelectrolyte effect. The following results are worth to be mentioned (Fig. 1).

In the case of PVA, while ΔG^0 varies continuously with α, ΔH^0 remains constant in a large range of pH's ($\alpha = 0.5–0$). This means that only the entropy term ΔS^0 is responsible for the pK variation in the above range. A comparison with a simple tetraamine (triethylenetetramine) shows that all the thermodynamic values for PVA are similar to those obtained with simple low molecular weight amines.

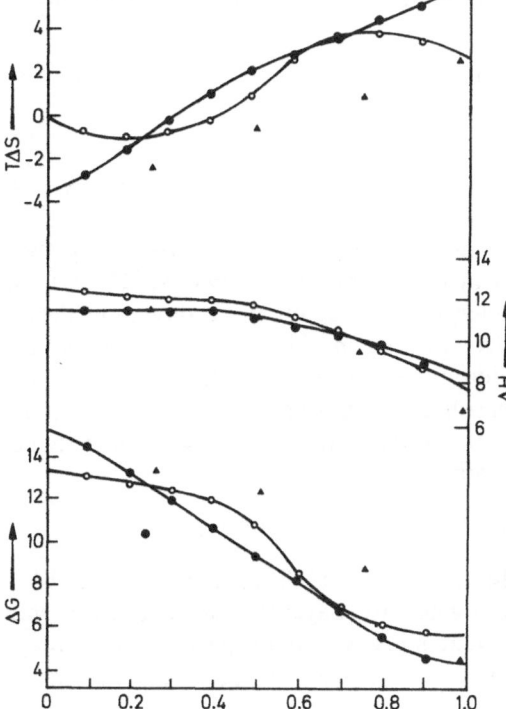

Fig. 1. Thermodynamic parameters (kcal/mol) for dissociation of PVA · HCl (●) and PEI · HCl (○), respectively. PVA · HCl at 0.1 M in 1.0 M KCl was titrated with 0.1 M KOH and PEI · HCl at 0.01 M in 0.9 M NaCl was titrated with 0.10 M NaOH. For comparison, the titration of triethylenetetramine tetrahydrochloride (▲) is included

Furthermore, the variation with α of the thermodynamic functions show the same trend in polymer and model, but they coincide in the two cases only at low charge densities. At pH values $\simeq 7$, the formation of a stiff structure is hypotised, involving the formation of relatively stable 6-membered rings along the macromolecular backbone, through hydrogen bonds between neighbouring free and charged aminic groups:

The above hypothesis is confirmed by ^{13}C and ^{15}N nmr data[58], and is similar to other hypotheses made in the case of simpler amines [59].

The variation of ΔH^0 with α, for α values >0.6, is obviously in relation with the increasing of the acidity of the $-NH_3^+$ ion, when more and more surrounded by positively charged groups.

In the case of PEI, the trend of ΔH^0 with α is very similar to that of PVA, but a plot of ΔG^0 versus α shows a sigmoidal shape, suggesting a conformational variation due to long-range effects [10].

Structurally related to PEI is a family of synthetic polymers having general formula:

$$\left[(CH_2)_x - \underset{\underset{CH_3}{|}}{N} - (CH_2)_y - \underset{\underset{CH_3}{|}}{N} \right]_n$$

where x and y may have the same value (2,3 or 6), or different values (x = 4; y = 3 or 6; x = 6; y = 3 or 5).

Since the above polymers are water insoluble in the form of free bases, their protonation behavior has been studied in mixed solvents (H_2O/CH_3OH 1:1). As expected, it was found that the dependence of pK_{app} on the protonation degree is stronger when x and y are smaller, i.e., when the charge density on the macromolecular chain at high protonation degrees is higher [22]. In other words, the "polyelectrolyte" character tends to vanish by increasing the length of the polymethylenic chains between the tertiary amino groups. When x = y = 6, the polymer seems to have the same behaviour as a monoamine (Fig. 2).

3.1.2 Heavy Metal Ion Complexing Ability

The heavy metal ion complexing abilities of PEI and PVA, and of some PVA derivatives, have been thoroughly investigated by Bayer [60]. No complexes with clear-cut stoicheiometries could be isolated; moreover, the capacities were always significantly lower than those calculated on the basis of the total amount of amino groups present.

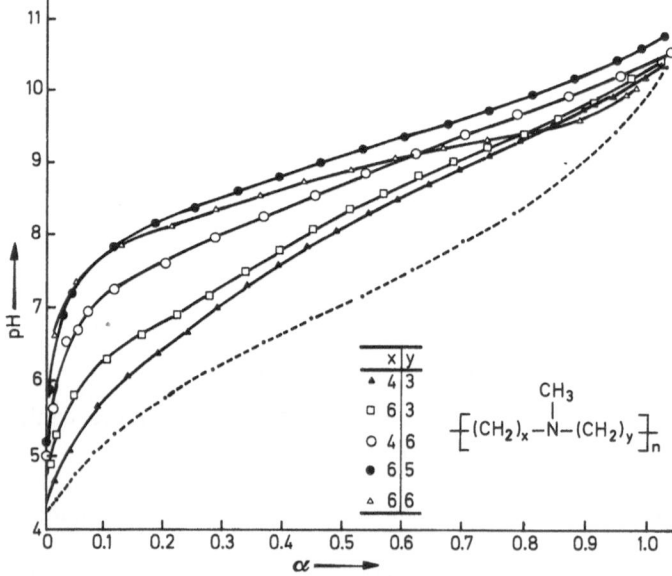

Fig. 2. Titration curves of linear poly(tertiary amine)s as hydrochlorides with different (x, y) values. (α = neutralization degree)

In the case of PVA, spectrophotometric and potentiometric titrations showed that two kinds of Cu^{2+} complexes may be present in aqueous solution: a "simple" complex, Cu-PVA, and a hydroxo complex. The latter occurs at pH values ≥ 7. In the Cu-PVA complex the Cu^{2+} ion seems to coordinate four $-NH_2$ groups. As a consequence, the shape of the macromolecule is deeply modified upon chelation, as confirmed by conductometric and viscosimetric measurements [61]. The initial viscosity of the PVA solution, in the presence of Cu^{2+} ions is reduced by a factor of about 10 at a pH corresponding to about 50% chelation. The same results are obtained in the presence of Ni^{2+} and Zn^{2+} ions [61].

3.1.3 Cross-linked Resins

A titration curve determined in 0.03 M NaCl at room temperature has been reported in the case of cross-linked PVA [62]. The shape of the curve is similar to that of the linear polymer. Cross-linked resins of polyethylene imine structure have been used to complex heavy metal ions from dilute solutions. The complex formation of these resins with metal ions was not thoroughly studied on a thermodynamic basis; however, the authors [63] demonstrated that, from a practical point of view, at least one of these resins in column operations was able to concentrate Cu^{2+}, Co^{2+} and Ni^{2+} from aqueous solutions, even in the presence of high concentrations of alkali- and alkaline earth metals.

3.2 PTN

3.2.1 Protonation and Complex Formation

Most polymers belonging to this class are water-soluble only in their protonated form. As a consequence, most potentiometric titrations of these polymers have been performed with OH^- starting from their hydrochlorides, either in water or in mixed solvents, such as water/dioxane, water/sulfolane, water/acetonitrile [64]. Many poly[thio-1-(N,N-dialkyl-aminomethyl)ethylene]s are optically active, and in these cases ORD and CD techniques may be used to study their protonation behaviour, in addition to potentiometric techniques.

The pK_a values are "apparent", and they depend on the solvent system in which they have been measured. The usual Henderson-Hasselbach expression of pK_a as a function of α is valid for each solvent system, as far as the reaction mixture is homogeneous. In those cases in which, at a certain pH, the reaction mixture becomes heterogeneous, the following expression has been found to be valid [65]:

$$pH - pH_p = 1/\langle P \rangle \int_{cM_{ppt}}^{cM} 1/(1 - \bar{\alpha}^*) \, d \log cM$$

were pH_P is the pH at the beginning of precipitation, $\langle P \rangle$ the average polymerization degree, cM the molar concentration of polymer in term of monomeric unit still in solution, and $(1 - \alpha^*)$ the average number of protonated sites in solution (Fig. 3).

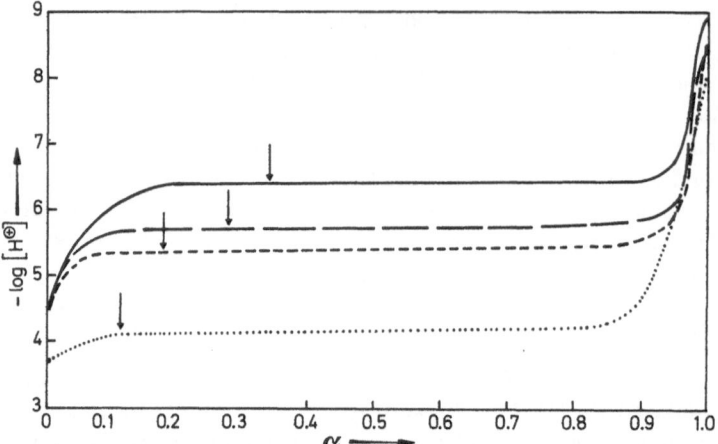

Fig. 3. Neutralization curves of the polyhydrochlorides of different PTN, namely poly[thio-1-(diethyl-aminomethyl)-ethylene hydrochloride] ————; poly[thio-1-(N-sec-butyl,N-methyl-aminomethyl)ethylene hydrochloride] — — —; poly[thio-1-(N-sec-butyl,N-ethyl-aminomethyl)ethylene hydrochloride] -----; poly[thio-1-(N-methyl,N-(1-phenylethyl)aminomethyl)ethylene hydrochloride] .... (α = degree of neutralization). (c = $5 \cdot 10^{-3}$ M dm^{-3} in 50/50 0.05 M aqueous KCL dioxane solution). The beginning of the gelation of partially neutralized polyamines is indicated by arrows. The smaller is the basicity, the smaller is the degree of neutralization by KOH at which the gelation appears

The rotatory power, and circular dichroism, have been found to have a linear dependence on $\bar{\alpha}$ [66]. This indicates that no conformational variations occur, contrary to what happens in the case of ionizable polypeptides [67], a fact that may be explained by the low degree of stereoregularity of these polymers, which cannot be expected to induce preferential conformations in solution. This is in agreement with the constance of pK_a in the condition used [68], i.e., in the presence of salt, while a conformational transition should result in a pK_a variation.

In the three optically active poly(amines) studied,

the protonation of the tertiary amino groups strongly affects the electronic transition n → σ* of nitrogen, and the related Cotton effect, which both disappear when the non-bonding electrons are captured, because involved in the protonation process.

These conclusions have been reached by comparing the CD or and ORD curves, and uv spectra of the polymers as free bases with those obtained for the same polymers in their protonated form [29].

A weak point of this study is that the authors were compelled to use two different solvents for the free bases (organic solvent), and for their hydrochlorides (water) [68].

The polymers (+)poly[thio-1-(N-sec butyl,N-methyl-aminomethyl)ethylene] and (−)poly[thio-1-(N-methyl,N-(1-phenylethyl)aminomethyl)ethylene] seem to be able to give complexes with Cu^{2+} in dioxane/ethanol 1:9. The nature of these complexes is not yet fully established; however, an interaction between Cu^{2+} ions and sulphur atoms probably occurs [29].

Partially quaternized poly[thio-1-(N-R_1,N-R_2-aminomethyl)ethylene]s are water-soluble. This allows to study their acid-base behaviour in water over the whole protonation range [69].

By adding acid to the above polymers in their free-base form, the following reactions take place:

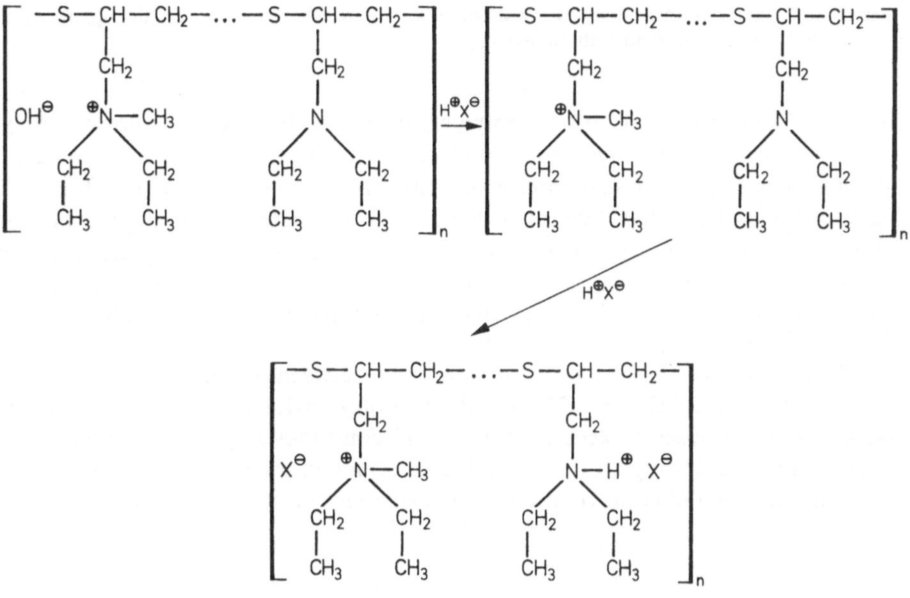

The first one involves an anion exchange between OH^- and X^-, and has little influence on the conformational arrangement of the polymer; the second one involves the protonation of uncharged tertiary amino nitrogens, and, as a consequence, the net charge of the polymer increases. This is confirmed by viscosimetric measurements in water, showing that no variations of $\eta sp/c$ occur after the first additions of H^+ [70]. On the contrary, the reduced viscosity sharply increases when the tertiary amino nitrogens start to protonate and the behaviour of the polymer is similar to that generally observed in polyelectrolytes (Fig. 4).

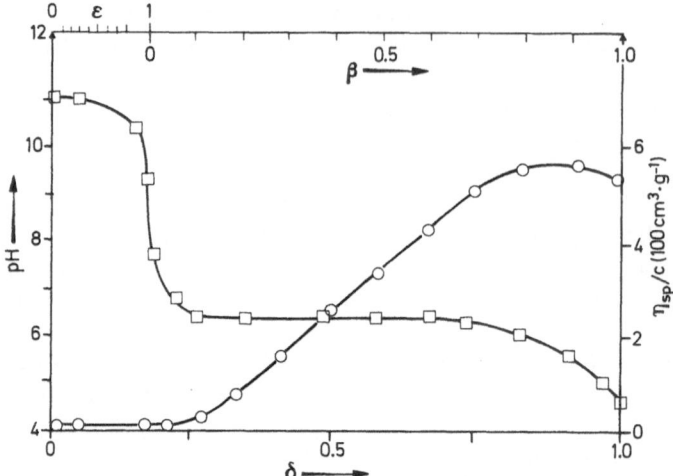

Fig. 4. Potentiometric –□– and viscosimetric –○– titration curves using HCl as titrating reagent (β = degree of protonation of residual tertiary amine groups; δ = overall degree of neutralization; ε = degree of neutralization of quaternary ammonium hydroxide sites) of partially quaternized poly[thio-1-(N,N-dimethyl-aminomethyl)ethylene]

Since the reduced viscosity of the polymer, in its free-base form, is considerably lower than expected on the basis of its molecular weight, the Authors contend that the polymer has a globular structure, which uncoils on protonation (Fig. 5).

The variation of the reduced viscosity values with the neutralization degree is lower when the quaternization degree is higher. The titration curves show a plateau whose width diminishes by increasing the quaternization degree. A similar plateau is also present in the titration curves of the unquaternized polymer in water, and in this case corresponds to the precipitation range (Fig. 6).

A model, based on hydrophobic interactions between tertiary amino groups, and on the lability of the $N-N^+$ bond of protonated weakly basic amines in water, has been proposed to account for the pH constancy in spite of the apparent water solubility. According to this model, deprotonated macromolecules assume a highly compact globular conformation with permanent positive electric charges, due

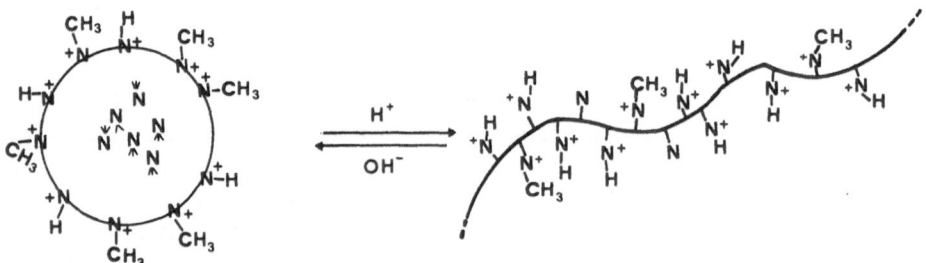

Fig. 5. Representation of macromolecules in the globular conformation and in the highly protonated extended state

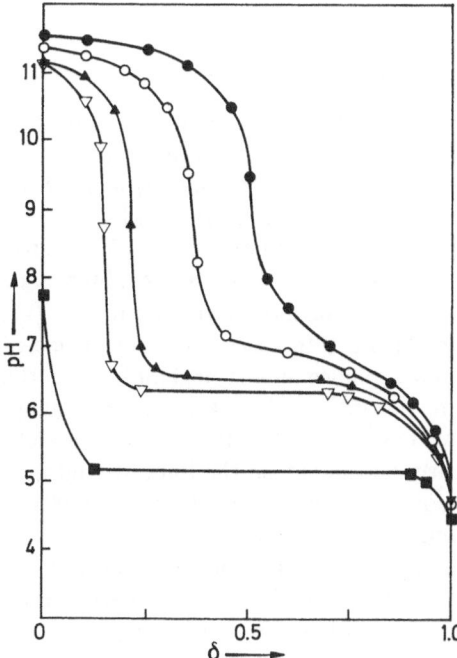

Fig. 6. Potentiometric titration cuves of dibiasic 17 (▽); 22 (▲); 35 (⊖) and 48 (●) and of partially quaternized poly[thio-1-N,N-dimethyl-aminomethyl)ethylene] (⊟) in water [titrating reagent, 1 M HCl)]. The numbers refer to the percent of quaternized nitrogens

to quaternary ammonium sites, arranged at the surface. The inner part of these globules constitute a water-free organic microphase wherein hydrophobic segments of deprotonated tertiary aminic repeating units stick together. In the highly buffered region, deprotonated macromolecules located in the microphase, which is equivalent to a precipitated phase in so far as protonation of tertiary amino groups is concerned, are at equilibrium with protonated and solvated ones in the solution phase. The actual degree of protonation of macromolecules in the solution phase, which depends on the polymer concentration, is much higher than the value the same macromolecules would have in a conventional one-phase system. Accordingly, the addition of acid or base to the system in the high-buffering region changes the distribution of macromolecules between the two phases. This exchange, which is equivalent to a cooperative transition from a globular deprotonated state to a highly protonated extended one, is believed to proceed through an all-or-none mechanism [71]. ORD and CD studies on the partially quaternized optically active poly[thio-1-(N,N-diethylamino-methyl)ethylene] containing 17% of N-methylated repeating units [Q-P(TDAE)*17] show, however, that the chiroptical properties changes depend on the content of both neutral, and charged (methylated of protonated) units, whatsoever the conformational state. This feature argues in favour of a random conformation of the macromolecules in the globular state, though their main chains were configurationally enriched.

3.3 P4VP and P2VP

3.3.1 Protonation

Both poly 4- and 2-vinyl-pyridine are water insoluble as free bases, and their protonation behaviours can be studied more easily in mixed solvents such as, for instance, ethanol/water mixtures. In this solvent system (45% ethanol) it has been found that the basicities of the polymers are much lower than those of their non macromolecular models, such as for instance 4-ethylpyridine (pK_a = 3.3 vs. 5.04) or 2-ethylpyridine (pK_a = 3.45 vs. 5.02). The pk_{app} of P4VP in the absence of added salt decreases very sharply by increasing the ionization degree up to α = 0.1, then it remains approximately constant (Fig. 7). In both cases, the dependence of pk_a on α is lower in the presence of NaCl, and becomes almost negligible when the NaCl concentration is 5×10^{-2} M/l. In the meantime the pK_0 values increase by increasing the concentration of NaCl, but never approach those of the models. The pK_0 values of the models are only slightly dependent on the ionic strength, as usually found for nonmacromolecular compounds (Table 3). In the case of P2VP, the reduction of pk_{app} with α is more pronounced than in the case of P4VP, and no plateau is present. Furthermore, the addition of NaCl does not induce a constancy of pk_{app} with α, even if the slope of the relative curve is significantly reduced at high salt concentration (Fig. 7). Viscosimetric measurements show that the η_{spec}/c values of the polymers (molecular weight of P4VP = $1.8 \cdot 10^5$, of P2VP = $1.4 \cdot 10^5$) sharply

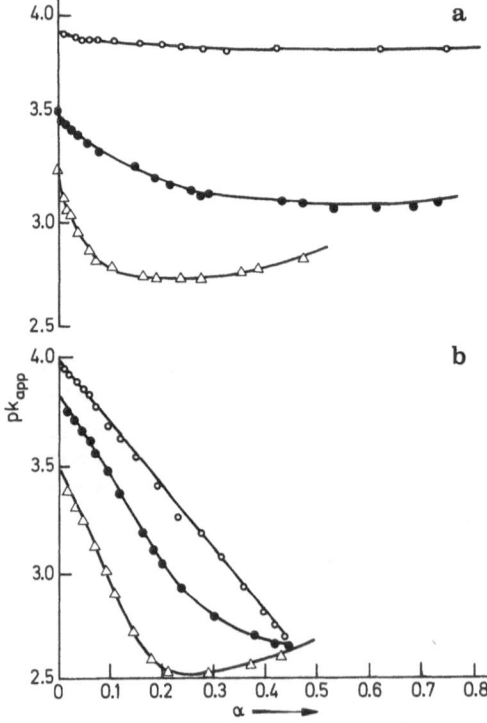

Fig. 7a and b. Dependence of pK_{app} for P4VP (**a**) and P2VP (**b**) on the degree of protonation at different ionic strengths of NaCl: $\triangle$—$\triangle$ without NaCl; $\bullet$—$\bullet$ 10^{-2} M/l; $\bigcirc$—$\bigcirc$ 5×10^{-2} M/l

Table 3. pK_0 of P4VP and P2VP and pK_a of their Monomer Analogues $4EP_y$, $2EP_y$, at 25 °C

Salt NaCl(M/1)	pK_0 in mixture[a]		pK_a in mixture[a]		pK_a in water	
	P4VP	P2VP	$4EP_y$	$2EP_y$	$4EP_y$	$2EP_y$
0	3.25	3.45	5.04	5.02	6.1	6.12
10^2	3.55	3.80	5.00	5.00	—	—
5×10^2	3.95	3.95	5.02	5.00	—	—

[a] 45% by weight ethanol and 55% by weight water, $pK_0 \pm 0.04$

increase by increasing α, till a value of about 0.2, and then decrease. As expected, this effect is much less pronounced in the presence of salt. The fact that the pK_0 values of the polymers are much lower than those of their models cannot be explained by electronic effects, since ^{13}C nmr spectra show that the chemical shifts of the aromatic carbon nuclei are practically the same in both cases. The authors [72] contend that a possible explanation of the low pK_0 of the polymers lies in the low values of the effective local dielectric constant [73] near the nitrogen atoms of pyridine residues, which are embedded in a hydrophobic environment due to the coiling of the macromolecule. On the other hand, the marked dependence of η_{spec}/c and pK_{app} on α may be caused by long-range electrostatic interactions between positive charges. Such interactions can be understood only if the electrostatic fields are thought to act mainly through the nonpolar organic part of the macromolecule. When neutralized by an acid, P2VP, insoluble when unionized, is protonated according to:

$$P2VP + H_3O^+ \rightleftharpoons P2VPH^+ + H_2O$$

and behaves as a cationic polyelectrolyte. The degree of protonation is lower than the theoretical degree of neutralization. As a consequence, a part of the added acid, which has not reacted with the polybase, remains in solution. The presence of this free acid acts much the same way as an added salt, and explains most, if not all, of the properties of P2VP. Due to the low basicity of pyridinic groups, it was difficult to obtain the ionization degree of P2VP from its potentiometric titration. The Raman spectra indicate that a conformational transition occurs in the polymer at $\alpha = 0.5$, while between 0.0 and 0.5, the polymer is progressively solvated and extended [74]. It is assumed that each proton is being shared by two pyridine rings and both rings can be regarded as partially protonated and partially neutralized.

The dependence of the protonation degree on the neutralization degree has been determined for various acids from spectroscopic and conductivity measurements [75]. It was found that the protonation behaviour is only slightly dependent on the nature of the counterions, and on the nature of the macromolecular chain (the isotactic polymers are less protonated, but the difference is very small) while a large amount

of electrolyte increases the apparent basicity of the pyridinic group [76]. Some differences between the protonation behaviour of isotactic and atactic poly(2-vinyl-
pyridine) in aqueous solution have been dscribed by Putermann et al. [77]. The
viscosity increases with increasing neutralization until it reaches a maximum at
about $\alpha = 0.5$. From that point on, the viscosity decreases with increasing neutralization reaching at $\alpha = 1.0$ a low value similar to that of the polymer solution at
very low degrees of neutralization. Both the atactic and isotactic polymer solutions
exhibit, in general, a very similar pattern of viscosimetric behaviour, although the
absolute viscosities of the two species differ to a large extent (the molecular weights
were very close, $\simeq 150.000$ in both cases). It seems, therefore, that the much lower
viscosity observed for the isotactic P2VP results from a limited extension of that
polymer, extension which is much smaller than that of the atactic polymer over the
whole range of neutralization. This conclusion might be supported by assuming either
some local helical character of the isotactic polymers in solution, or that a crystalline
polymer may preserve some of its local order in solution [78]. It may be noted that
the contribution of non-electrostatic forces to the conformation of chains is of greater
importance in the case of isotactic configuration, favouring intrapolymer interactions
acting in an opposite direction to the repulsions of changes.

The position of the maximum in the reduced viscosity curves depends on the
degree of neutralization and on the nature of the counter-ion [76]. In the presence of
Br^- or NO_3^- the reduced viscosities are different from those in the presence of Cl^-;
this can be due to changes in the binding of counterions. [79]. The polymeric ion exhibits
a stronger affinity for NO_3^-, according to the results of Donnan equilibrium
measurements at $\bar{\alpha} = 0.5$ [80].

The effect of divalent ions such as SO_4^{--} has been studied either by electric permitti-
vity techniques, or by viscosimetry. It was experimentally observed that the dielectric
increment is larger than in the presence of monovalent ions. In the presence of the
bivalent counterions there are two opposite effects. From one hand, the bivalent
ions tend to increase the dielectric increment through a charge effect, while from
the other hand the observed reduction of the viscosity proves that in the presence of
SO_4^{--} the polyions contract [76].

The trends of viscosity and dielectric parameters during the neutralization show
that the maximum extension of the polyions occurs when only a part of the ionizable
groups are neutralized.

In the presence of 0.1 M NaCl the viscosity of the various solutions remains at a
constant value throughout the whole range of neutralization [77].

Light scattering, turbidity and spectroscopic studies show a transition between
two conformational states of the chains in solution which are in a dynamic equilibrium [74].

Partially quaternized poly(4-vinylpyridine) is soluble in water, and can be studied
in aqueous solution [81].

On the whole, the results obtained by studying poly(4-vinylpyridine)s partially quaternized with benzylchloride (P4PyBz) or bromoacetone (P4PyAc) in water, are consistent with those obtained with unquaternized poly(4-vinylpyridine)s in mixed solvents. The pK_0 values of the partially quaternized polymers are only slightly dependent on the quaternization degree. To some extent, they also depend on the type of the quaternizing agent [81]. A simple electrostatic effect does not seem to play an important role in lowering the basicity of the polymer with respect to its model [$pK_0 = 2.4$(P4PyBz), and 3.3(P4PyAc); pK_a of 4-ethylpyridine $= 6.1$]. According to the authors, the most probable explanation of this fact lies in a microenvironmental effect. In other words, a partial dehydration of the microregion near to the nitrogen atoms probably occurs, thus intensifying the electrostatic interactions between the different units of the polymer. Random copolymers of the following structure:

(where $X + Y + Z = 100\%$), have been studied by viscosimetric and potentiometric techniques, together with similar derivatives of quaternized poly(4-vinylpyridine), in which the alkyl groups on nitrogen carried an NH_2 group in ω position [82].

The pK_0 in aqueous solution of polymers having the same overall quaternization degree ($\simeq 51\%$) show a marked dependence on the amount of dodecyl groups present in the macromolecule. For instance, when no dodecyl groups were present, the pK_0 was 4.4, while a sample in which about 17.5% of alkyl groups were dodecyl groups, the pK_0 was 3.7 (at 25 °C). Moreover, no major conformational transitions take place in these materials as a function of the ionisation degree, irrespective of whether the polymers under study were to be considered polyelectrolytes, or polysoaps. The positions of the potentially charged groups (free pyridine nitrogens) are such, that upon ionisation, they do not disrupt the hydrophobic effect operating through the nonpolar hydrocarbon chains. In the case of N-(ω-amino)alkyl derivatives, like the aminobutyl- and the amino dodecyl ónes, viscosity measurements show that a significant expansion of the polymeric chains takes place by increasing the ionisation degree of the amino groups. At $\alpha = 0$ these molecules seem to be relatively compact, due to pockets of aggregated hydrophobic groups; their micellar structures are stabilized by the charged quaternary nitrogens near the periphery of the pockets. The free amino groups are thought to be largely incorporated into the hydrophobic regions. Upon ionisation of the amino groups, the size of the hydrophobic pockets gradually diminishes, and the resulting additional hydration is accommodated for by a corresponding expansion of the polymer chain [82].

3.3.2 Heavy Metal Ions Complexing Ability

The stability constants of the Cu^{2+} complexes with poly(4-vinylpyridine), and with partially quaternized poly(4-vinylpyridine) have been determined by potentiometric techniques [83]. In some cases the Cu^{2+} ion seems to coordinate four pyridinic groups of the polymer, and the overall stability constant of the complex (log $\beta' = 10.8$) is significantly higher than that of Cu^{2+} complexes with non macromolecular 4-vinyl-pyridines (log $\beta' = 6.5$). This has been attributed to a polychelation effect by different units of the polymer. A partial quaternization has a very poor effect on complex formation, but this effect, though small, depends on the quaternizing agent. For instance, with ethyl bromide the overall formation constants decrease by increasing the quaternization degree, while with benzyl chloride the quaternization degree does not influence the stability of the complex. The viscosity of solutions of the Cu-PVPy complex in water decreases by increasing the pH, till pH 5.4. This suggests that the complex tends to assume a very compact structure as a whole. The viscosity increases again above pH 5.4, due to the hydrolysis of the complex.

By contrast, in the case of water solutions of a poly(4-vinylpyridine) partly quaternized with benzylchloride, it was found that by plotting η_{sp}/c versus the [Cu]/[N] ratio (where N refers to unquaternized nitrogens), a constant value is reached at a ratio = 0.5, above which no viscosity changes are observed by further additions of Cu^{2+} [84]. These data apparently suggest that the Cu^{2+} ions coordinate only two pyridine groups.

3.3.3 Cross linked Resins

Most studies on metal ions adsorption by cross-linked poly(vinylpyridine) resins have been performed on resins based on poly(4-vinylpyridine), which apparently gives more interesting results that its 2-analogue. Some metal ions are effectively adsorbed by these resins. According to the conditions, the sorption may take place either by anion exchange, or by complex formation with the pyridine moieties. The sorption of metal ions by the latter mechanism strongly depends on the surface properties of the resins. Poly(4-vinylpyridine) may be complexed with metal ions and then cross-linked with α,ω-dibromoalkanes, according to previously reported ideas of polymer cross linking in the presence of template [85]. This procedure is illustrated in Fig. 8, and involves essentially four steps: i) a complex between a polymeric ligand and the metal ion is first prepared in solution. ii) This complex is cross-linked by the addition of a suitable cross-linking agent to its solution. iii) The metal ion

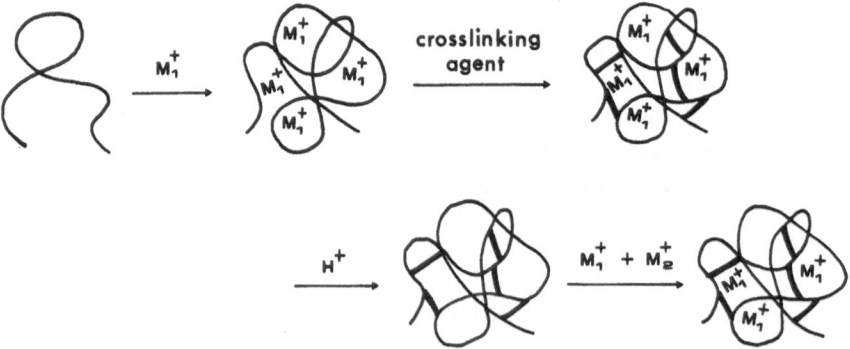

Fig. 8. Scheme showing binding of metal ions to polymer

is removed from the resulting resin by acids. iv) The resin is washed with dilute alkali and water [86].

If the conformation of the polymeric ligand chain is somewhat "fixed" by cross-linking, the resulting resin, when dipped into a solution containing various metal ions, will be selective for the metal ion to which it had been previously coordinated. This was borne in practice in the case of poly(4-vinylpyridine) resins, in which the cross-linking agent was 1,4-dibromobutane. The amounts of metal ions coordinated by the pyridine groups on the resin are given in Table 4, where the metal ion in parentheses indicates the metal ion used as template.

Branched and linear poly(ethylene imine), and poly(vinylimidazole), were similarly cross-linked in the presence of template Cu^{2+} or Co^{2+} ion, but no enhancement of selectivity towards the same metal ions was achieved [37].

It may be mentioned that some ion-exchange resins having a poly(4-vinylpyridine) matrix partly quaternized with ω-aminoalkyl groups have been prepared and studied with interesting results, as far as their thermal regenerability is concerned [87]. This is apparently a recent development of a wide research project on thermally regenerable ion-exchange resins [88].

Table 4. Adsorption of metal ions on crosslinked PVP resins[a]

Resin	Adsorbed metal ions, %			
	Cu^{2+}	Co^{2+}	Zn^{2+}	Cd^{2+}
Cross-linked (Cu)	52	6	8	9
Cross-linked (Co)	16	10	8	9
Cross-linked (Zn)	8	6	11	6
Cross-linked (Cd)	9	4	7	8
Cross-linked (PVP)	15	7	6	6

[a] $[\bar{N}] = 0.01$ M/l, [metal ion]/$[\bar{N}] = 0.1$, in $CH_3COOH—CH_3COONa$ buffer, pH 5.5, where $[\bar{N}]$ denotes unquaternized and coordinatable vinylpyridine groups in the resin

3.4 PVI and P4VI

3.4.1 Protonation and Complex Formation

Poly(N-vinyl-imidazole) is an essentially monofunctional polybase, as shown by the sharpness of the jump in the titration curve [89]. From conductance measurements, it was found that the data fitted the Henderson-Hasselbach expression with a term for the ionic strength:

$$p(c\alpha) = pk_{av} - n \log \left(\frac{1-\alpha}{\alpha}\right) - \Phi \sqrt{c\alpha}$$

where $p(c\alpha)$ is the negative logarithm of the counterion concentration, and the ionic strength is set proportional to the ionic concentration of the polysalt solution;

Table 5. Parameters of Henderson-Hasselbalch Equations in PVI Salts
with different acids

Counterion	n	pK_{av}	Φ
Iodate	2	3.00	13.6
Chloride	2	3.20	15.4
Bromide	2	3.36	16.0
Nitrate	2	3.36	16.0
Iodide	2	4.36	24.1
Trichloro acetate	2	4.10	23.8
p-Toluene-sulfonate	2	4.18	23.0

pK_{av} are average pK values; Φ = corrective term

n was found equal to 2 for all the polysalts at all concentrations. The average pK values (pK_{av}) and the corrective term (Φ) are given in Table 5 for different counterions. The relative constancy of Φ for all but one of the small anions suggests that the same statistical unit is involved. The large value of Φ with the large anions was suggested to arise from the fact that these anions are largely located outside the polymer coil, while the smaller anions may be largely inside the coil.

The complexing ability of the PVI has been studied with several heavy metal ions [89,90]. The stability constants of the complexes have been determined in aqueous solution by the method of Bjerrum [91], modified for the binding of metals by polyelectrolytes [92]. It could be concluded that both Cu^{2+} and Zn^{2+} ions coordinate four imidazolyl groups. (Table 6).

Table 6. Stability Constants of Cu(II) and Zn(II) complexes with PVI in 1.0 M $NaNO_3$ at 24 °C

	$\log K_1$	$\log K_2$	$\log K_3$	$\log K_4$
PVI (0.01 M) — Cu^{2+}	3.50	3.42	3.88	3.76
PVI (0.01 M) — Zn^{2+}	2.32	2.83	3.17	3.16

An increase of the successive stepwise formation constants is observed. This effect appears to be quite general with polymeric complexing systems: once the metal ion is coordinated to one group of the polyelectrolyte coil, the high local concentration of available ligands triggers the completion of binding, and make the apparent successive formation constants larger than the first. The overall constant increases with the ionic strenght, in a way more pronounced than in the case of the corresponding non-macromolecular models, owing to the effect of the electrolyte on the polymer's conformation.

In the case of Ag^+, only two imidazolyl groups participate in complex formation, and the overall stability constant is $\log \beta_2 = 8.00$, about one order of magnitude larger than the overall formation constant of the Ag/imidazole complex ($\log \beta_2 = 6.84$). Complexes of Cu^{2+} with poly 4(5)-vinyl-imidazole, poly(N-vinyl-imidazole), and poly(2-methyl-N-vinyl-imidazole) have been studied in aqueous solution by Sato, Kando, and Takemoto by spectroscopic techniques (visible, and esr)[93]. All the complexes seem to have a tetragonally distorted octahedral structure. All the variations observed by changing the pH, or the Cu^{2+}/polymer ratio, may be ascribed to distortions of the tetragonal structure, due to the very nature of the polymer, which does not favour the coordination of four imidazolyl groups around the Cu^{2+} ion. At high pH values, the spectra of the Cu^{2+}/polymer complexes become similar to those of the corresponding complexes with low-molecular weight models, while the viscosities of their solutions become lower, suggesting a contraction of their effective volumes.

By changing the nature of the anion, the visible and esr spectra show only slight changes. In all cases, the spectra of the Cu^{2+}/polymer complexes are different from those of their non-macromolecular analogues[94].

Copolymers having pendant sulfide and imidazolyl groups (copoly VS-4VI)

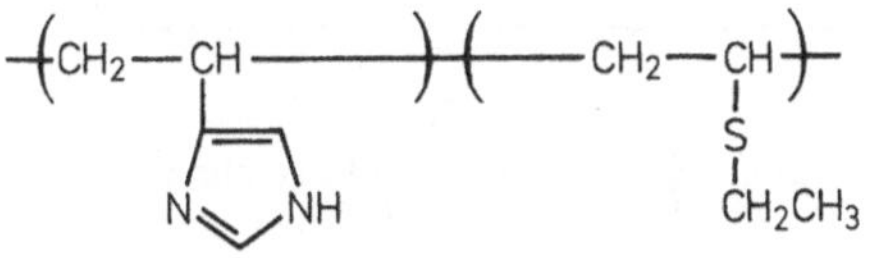

have been studied as ligands for both Cu^{2+} and Cu^{+}[95]. With respect to P4VI, it has been found that by increasing the VS residues in the Cu^{2+}/copoly(VS-4VI) complex, the optical density in the visible spectra slightly increased at first, and then tended to decrease. The absorption maximum in the visible region was shifted from 630 nm to 620 nm and then gradually to 690 nm. In the esr spectra, only small changes were observed when the VS component increased from 0 to 57%, but large changes when it increased up to 71%. These results, and even more the fact that the visible and esr spectra of Cu^{2+} P4VI complexes at various molar ratios of P4VI to Cu^{2+} were similar to those of Cu^{2+}/copoly VS-4VI complexes as far as the concentrations of the imidazolyl group was kept to be identical, indicate that the imidazolyl groups in the copolymers strongly coordinated the Cu^{2+} ions, and that the sulphur atoms of the sulphide groups were coordinated only loosely. The same groups, however, seem to be strongly coordinated to Cu^{+}, and this explains their catalytic power in the oxidation of hydroquinone.

3.5 PAA

The protonation and complex formation of a number of poly(amido-amine)s in aqueous solution have been studied by potentiometric, calorimetric, viscosimetric, spectrophotometric, esr, ^{13}C nmr, and quantum chemical techniques.

3.5.1 Protonation Studies

In all the poly(amido-amine)s studied, the results of the potentiometric titrations

clearly show that the basicity of the aminic nitrogens $-\overset{\mathclap{R}}{\underset{\mathclap{|}}{N}}-$ of each repeating unit does not depend on the degree of protonation of the whole macromolecule. Consequently, "real" basicity constants can be determined [96]. The number of basicity constants is in all cases equal to the number of the aminic nitrogens present in the repeating unit. On this respect, the behaviour of these polymers is very similar to that of their non macromolecular models, apart for some minor differences in basicity due to entropy effects. Calorimetric studies, in fact, demonstrate that the protonation enthalpies are always about the same for polymers and models, and a linear relationship can be obtained between the enthalpies and the net charges of nitrogens that undergo protonation [43]: Thus, it may be reasonably concluded that the inductive effect alone can grossly explain the differences in the enthalpy changes between different polymers. Plot of η_{spec}/c versus α gave different results for poly(amido-amine)s belonging to different classes (Table 7) [97]. When the experiments were per-

Table 7. Poly(amido-amines) and their models

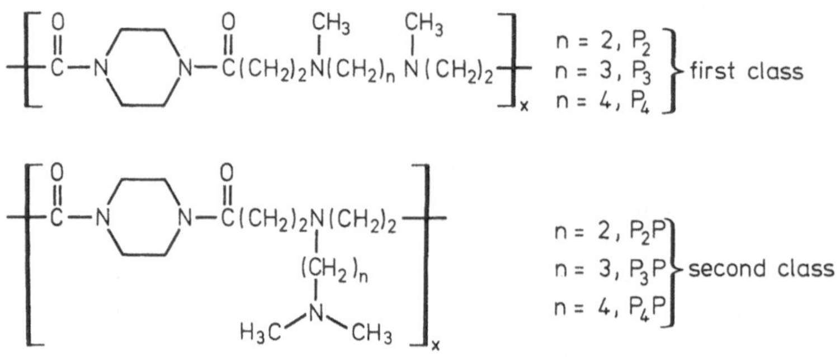

formed in 0.1 M NaCl, in the case of PAA having two aminic nitrogens in the repeating units, both of them belonging to the main chain, the η_{sp}/c versus α plots show two jumps, corresponding to the neutralisation points of the two basic nitrogens. On the contrary, in the case of polymers belonging to the second class, i.e., in which one of the aminic nitrogens was present as side substituent, the corresponding plots do not show critical variations of η_{sp}/c in the whole range of α. In other words, the viscosities of the first class of polymers are constant at low charge density, then they suddenly increase when the charge density reaches a value corresponding to the first equivalent point, and remain constant till the proximity of the second equivalent point, when they again suddenly increase, (Fig. 9). Both ^{13}C nmr results [42,97,98], and quantum chemical computations [97,99] show that no significant interactions exist between groups belonging to different monomeric units. Therefore, the conformational transitions induced by protonation take place within each monomeric unit. As far as the first class of polymers is concerned, the first protonation leads to the formation of a strong hydrogen bond, between "onium" ions and carbonyl groups belonging to the same monomeric unit. The second protonation leads to an electrostatic repulsion between the positively charged "onium" ions belonging to the same monomeric unit, and this effect compels the polymer in a rigid, more extended conformation. In the polymers belonging to the second

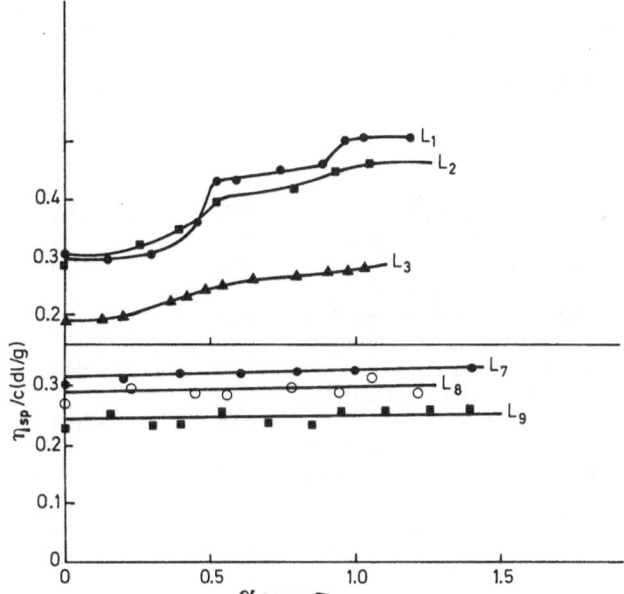

Fig. 9. Viscosimetric titrations of PPA's in 0.1 M NaCl

class, according to both ^{13}C nmr [97] and the potentiometric data, the first protonation occurs on the side-nitrogens [100]. As a consequence, no effective hydrogen bridges can be formed between the very distant "onium" ions, and the carbonyl groups. The second protonation step involves the aminic nitrogen of the main chain. The presence in the main chain of only one "onium" ion, which can interact indifferently with two neighbouring carbonyl groups, leads to a greater conformational freedom, hence viscosimetric measurements over the whole range of pH's do not show any "jump" even after the second protonation step.

3.5.2 Complex Formation

Most compounds listed in Table 1 give complexes with Cu^{2+} and Ni^{2+} ions in aqueous solution [96]. The only exceptions are the polymer with only one aminic nitrogen in the repeating unit, and the polymer with two aminic nitrogens belonging to a piperazine ring, together with their models. These results are not unexpected, because it is well known that tertiary mono-amines and N,N' disubstituted piperazines usually do not give stable complexes in aqueous solution. The fact that no complexes have been obtained also in the case of polymers is a further hint that no cooperative effects between nitrogens belongings to different units take place in complex formation.

Complexes with general formula CuL and NiL (where in the case of the polymeric ligands "L" means the repeating unit of the polymer) were evidenced in aqueous solution [96]. In some cases, hydroxylated and protonated species were also found [47, 100, 101]. The stability constants for CuL complexes are reported in Table 8.

As in the case of the protonation constants, the stability constants of the models are slightly higher than those of the polymers.

Table 8. Spectrophotometric data and stability constants of copper(II) complexes with poly(amidoamines) (Pm, Pn, P) and their non macromolecular models (Mn, MnP) at 25 °C in 0.1 M NaCl

Ligand (L)	Ability to form Cu^{2+} complexes in aqueous solution	Reaction	Stability Constants	Electronic spectra[a] 10^3 cm^{-1}
P_2^c	Yes	$Cu^{2+} + L \rightleftarrows CuL^{2+}$	8.96	14.8(174)
		$CuL^{2+} + OH^- \rightleftarrows Cu(OH)L^+$	5.52	
P_3	Yes	$Cu^{2+} + L \rightleftarrows CuL^{2+}$	5.36	14.9(169)
		$CuL^{2+} + OH^- \rightleftarrows Cu(OH)L^+$	5.14	
		$CuL^{2+} + 2 OH^- \rightleftarrows Cu(OH)_2L$	10.25	
P_4	NO	—	—	—
M_2^c	Yes	$CuL^{2+} + L \rightleftarrows CuL^{2+}$	9.10	14.8(176)
		$CuL^{2+} + 2 OH^- \rightleftarrows Cu(OH)_2L$	8.40	
M_3	Yes	$Cu^{2+} + L \rightleftarrows CuL^{2+}$	6.45	14.9(169)
		$CuL^2 + OH^- \rightleftarrows Cu(OH)L^+$	5.09	
		$CuL^{2+} + 2 OH^- \rightleftarrows Cu(OH)_2L$	9.90	
M_4	NO	—	—	—
P_2P	Yes	$Cu^{2+} + L \rightleftarrows CuL^{2+}$	8.47	14.2(135)
		$CuL^{2+} + OH^- \rightleftarrows Cu(OH)L^+$	6.12	
P_3P	NO	—	—	—
P_4P	NO	—	—	—
M_2P	Yes	$Cu^{2+} + L \rightleftarrows CuL^{2+}$	8.61	14.1(125)
		$CuL^{2+} + OH^- \rightleftarrows Cu(OH)L^+$	5.7	
		$CuL^{2+} + 2 OH^- \rightleftarrows Cu(OH)_2L$	8.7	
M_3P	NO	—	—	—
M_4P	NO	—	—	—

[a] The molar absorption coefficients (dm^3 mol^{-1} cm^{-1}) are given in parentheses

The electronic and esr spectra of the CuL complexes are consistent with an octahedral tetragonally distorted structure [101], as present in most Cu^{2+} complex compounds in aqueous solution. The spectroscopic measurements indicate that there are some contribution from the highly polarizable C=O groups. On the other hand, the spectra of the Cu^{2+} complexes with polymers have been found to be almost identical to those of complexes with the corresponding models [100, 101]. This suggests that for each polymer-model pair, the structure of the complexes is the same. Hence, as in protonation, the difference between the stability constants must be due mainly to entropic effects.

By comparing the stability constants relative to different polymers, and models, it may be observed that they increase with the number of basic nitrogens present in the repeating unit. Furthermore, the stability constants of the Cu^{2+} complexes of the polymers of the first class are higher than those of the isomeric polymers of the second class. This fact, and the lower d-d energy band of the complexes of the second class, has been explained with the lack of partecipation in the latter of the C=O groups to the metal coordination [102]. Viscosity measurements show that for the complexes with polymers of the first class, the viscosity monotonously decreases upon increasing the pH until the formation of the complex CuL is complete, and then remain nearly constant. On the contrary, in the case of the polymers of the second class, η_{sp}/c is

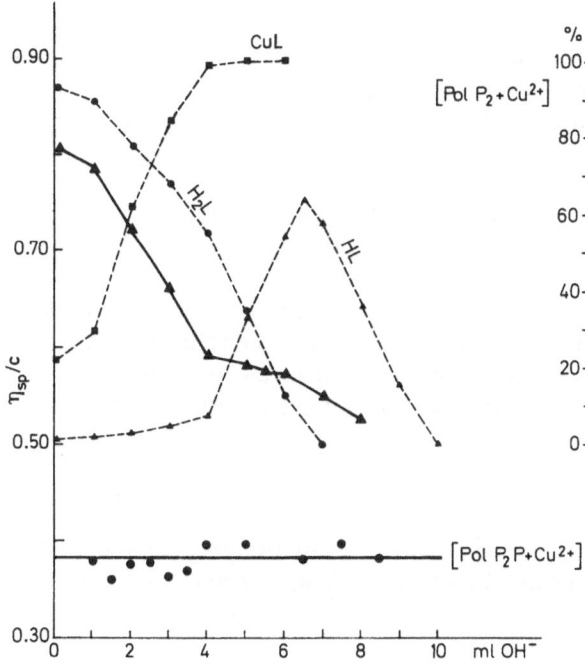

Fig. 10. Variation of the viscosity η_{sp}/c (solid lines) and distribution curves (broken lines) vs. millilitres of OH^- added for the two systems CuP2 and Cu-P2P

independent of the pH. This behaviour has been explained by the fact that only in the first case the conformational transitions involve the main polymeric chain, i.e., the whole macromolecule, while in the case of "unsymmetric" polymers the coordination only involves side groups (Fig. 10).

In the case of a polymer, and model, containing three nitrogen atoms:

$$\left[-CH_2CH_2CON\bigcirc NCOCH_2CH_2N\underset{CH_3}{|}\ CH_2CH_2N\underset{CH_3}{|}\ CH_2CH_2N\underset{CH_3}{|}-\right]_n \qquad P_2(3N)$$

$$O\bigcirc NCOCH_2CH_2N\underset{CH_3}{|}\ CH_2CH_2N\underset{CH_3}{|}\ CH_2CH_2N\underset{CH_3}{|}\ CH_2CH_2CON\bigcirc O \qquad M_{2(3N)}$$

the stability constants increase by about four orders of magnitude [47].

The reaction: polymer + Ni^{2+} → complex, is very slow. This makes it cumbersone to determine the values of the stability constants with sufficient accuracy. Qualitatively, however, it has been found that the polymers and models able to form complexes with Cu^{2+} ions behave in the same way with Ni^{2+} ions [96]. In one case, i.e. polymer P_2 and model M_2, the stability constants of Ni complexes have been determined, and found to be 5.7 and 6.1 respectively. Also in this case, it may be noted that the constants obtained with the model are slightly higher than those obtained with the polymers. It may also be observed that the constants obtained with Ni^{2+} are considerably lower than those obtained with Cu^{2+} [96].

3.5.3 Solid Complex Compounds

Solid Cu^{2+} complexes have been prepared from polymers and models P_2, M_2 and $P_2(3N)$ [47]. They separated by mixing methanol solutions of Cu^{2+} nitrate. $3 H_2O$, and ligand in $1:1$ molar ratio, and cooling. Elemental analyses were consistent with the formula:

$$CuL(NO_3)_2 \cdot x\, H_2O \qquad \begin{array}{l} x = 3 \text{ for Polymer } P_2, P_2(3N) \\ x = 1 \text{ for Model} \quad M_2 \end{array}$$

The electronic spectra of the solid complexes are practically identical to those of the same complexes in aqueous solution [101], as reported in Table 8. This means that, contrary to all previously described complexes of metal ions with polymeric ligands, the poly(amido-amines) complex compounds have a well defined stoichiometry.

3.5.4 Cross-linked Resins

Insoluble resins having essentially the same structure of the previously described poly(amide-amine)s have been prepared by using 1.8-diamine octane or 1.4-diamine butane as cross-linking agents:

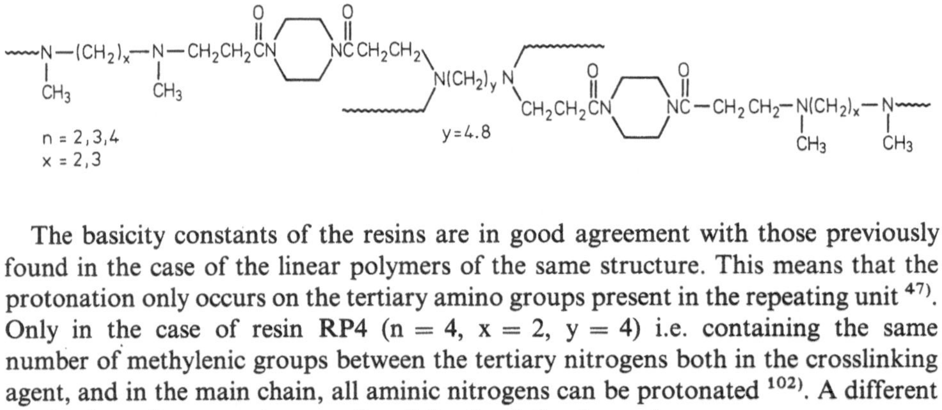

The basicity constants of the resins are in good agreement with those previously found in the case of the linear polymers of the same structure. This means that the protonation only occurs on the tertiary amino groups present in the repeating unit [47]. Only in the case of resin RP4 (n = 4, x = 2, y = 4) i.e. containing the same number of methylenic groups between the tertiary nitrogens both in the crosslinking agent, and in the main chain, all aminic nitrogens can be protonated [102]. A different mechanism of protonation was found for the following resin:

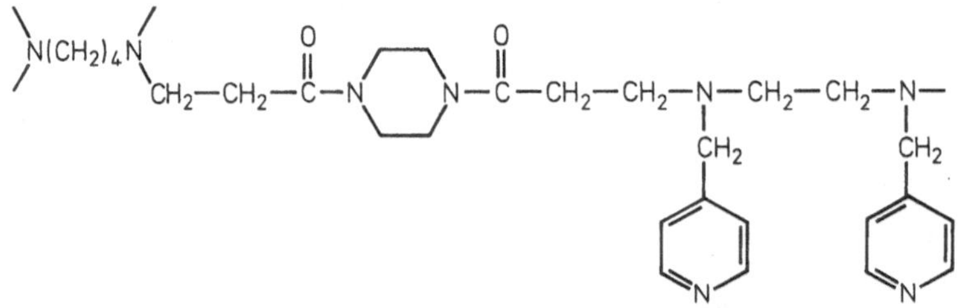

in which the first two constants are attributed to the nitrogen atoms of the crosslinking portion. This peculiar behaviour was attributed to the fact that the bulky substituents induce a very low degree of conformational freedom, resulting in a lower accessibility to the basic sites by the protons [103]. The complexing ability of the resins towards Cu^{2+} ions was studied potentiometrically at 25 °C in 0.1 M NaCl. The stability constants of the simple complex CuR (where R is the resin repeating unit) is always lower than that of the CuL complex (where L is the repeating unit of the corresponding linear polymer). The above difference seems to be due to the fact that the complex formation in the case of linear polymer occurs through a cyclic structure around the copper atom involving two C=O and two tertiary amino groups, and such a cyclic structure is obviously biased in the case of a tightly crosslinked resin [102]. In the case of the resin containing three tertiary amino groups, the stability constants of the CuR complex is in fair agreement with that obtained in the case of the linear polymer [47]. This means that the Cu^{2+} ion coordinates three tertiary nitrogen atoms also if the ligand is in a cross-linked form. Studies on the separation of several metal ions on a column of the same resins indicate that a sharp separation may be achieved between metal ions of otherwise similar behaviour in aqueous solution [104].

3.6 PMC

3.6.1 Protonation and Complex Formation

The complexation properties of various poly(diaza-crown ether)s have been studied either in water, or in organic solvents, according to their solubilities. Polymers prepared from epichlorohydrine (Epic) or diepoxyoctane (Epico) are water soluble. The structures of the repeating units of these polymers are depicted in Table 9.

These polymers do not exhibit a typical polyelectrolyte behaviour and "real" basicity constants can be calculated [105, 106]. According to the authors, this is due to the presence of the macrocycles, sheltering the positive charges on the protonated nitrogens, and to the low molecular weight of the samples (2000–5000 dalton, corresponding to 5–15 monomeric units). The basicity constants of the polymers are collected in Table 10, together with those of their non-macromolecular analogues. It can be seen that the polymeric structure has very little influence on the constants, as far as the first protonation is concerned, but decreases the basicity of the other aminic groups of the ring, except for P|22| Epic. This is attributed to the interaction between the first protonated group and the other aminic groups of the rings, which in the polymers P|22|Epic, P|21|Epic, and P|222$_N$|Epic, are in close proximity. This type of interaction is favoured by the polymeric nature of the ligands [107].

The stabilities of the complexes of alkaline- and alkaline earth cations with polymeric macrocyclic ligands, as well as with their non-macromolecular models, mostly depend on the relative sizes of the ring and the cations, the shape of which can be considered as spherical [108]. Other factors, however, affect both stabilities, and selectivities. The replacement of O by N in the rings decreases the electrostatic interaction of the cation with the ligand, and to some extent the size of the cavity. In addition, the chemical environment of the rings, and in particular the presence of neighbouring groups affect the stabilities of the complexes [109]. In principle, the neighbouring groups present in the polymers can perturb the electronic distribution of the donor atoms, change the hydration shell of the ligand, induce a steric or conformational strain, and also participate to the coordination of the

Table 9. Structure of some macrocyclic ligands

Segments binding	Macrocycles	Code names
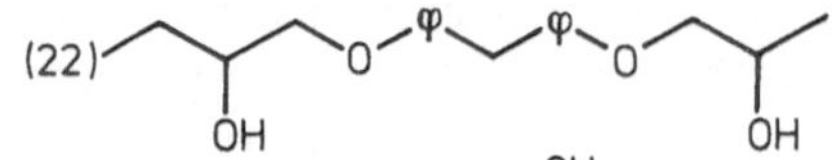 (22) [structure with O–φ–φ–O and OH groups]	[macrocycle (22) with N, O atoms]	p(22) Epi DPP
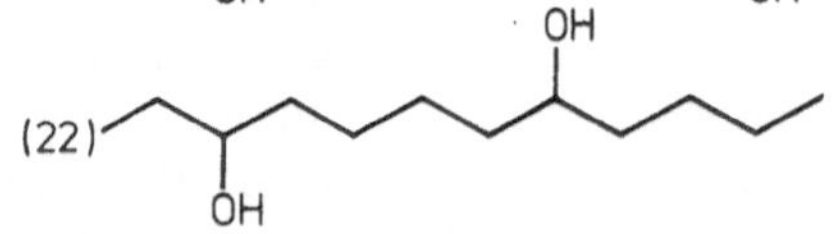 (22) [structure with OH groups]	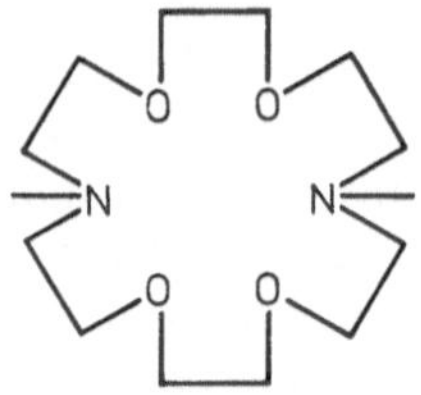	p(22) Epi I
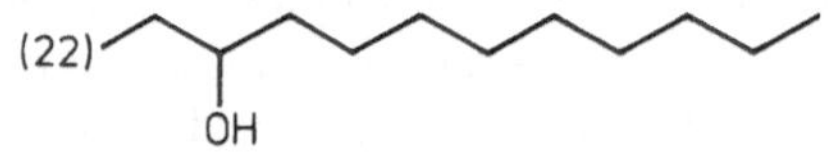 (22) [structure with OH group]	(22)	p(22) Epi C
 (22) [chain structure]	[macrocycle (222_N) with N, O atoms]	p(22) decan
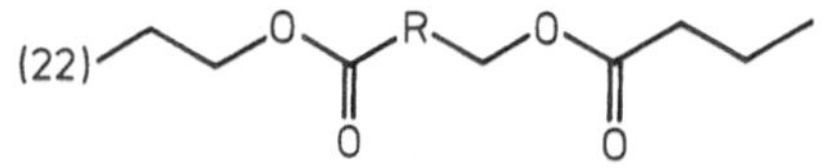 (22) [structure with O, R, O, carbonyls]	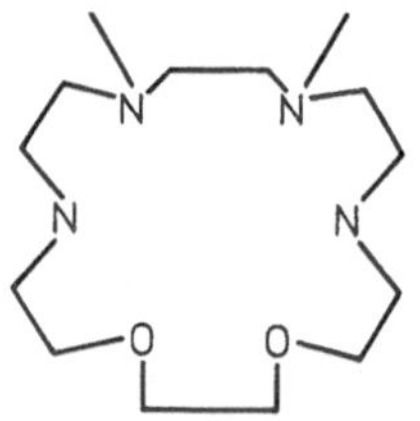	p(22) EtOH–R
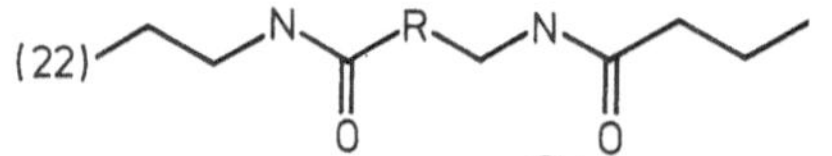 (22) [structure with N, R, N, carbonyls]	(222_N)	p(22) EtNH₂–R
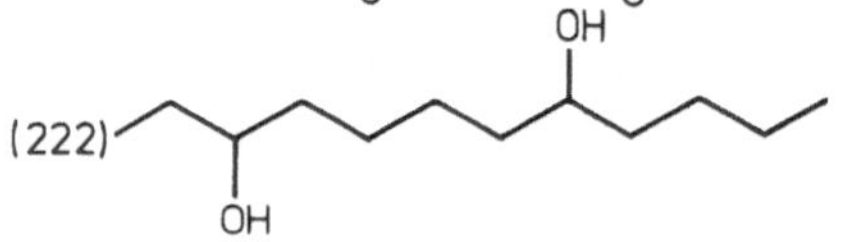 (222) [structure with OH groups]		p(222_N) Epi O

cations. However, comparison between the polymeric complexes, and the corresponding complexes with non-macromolecular models, show that no large differences exist. Only in the case of the complex of polymer P|22| Epio with Ba^{2+}, the polymeric nature of the ligand leads to additional properties. While Co^{2+}, Ni^{2+}, Cu^{2+}, Zn^{2+} cations are included in the macrocycle cavity, a 2:1 complex is observed with Ba^{2+}. An intramolecular "sandwich" structure is probably formed in this case, due to the fact that the size of the ring is too small to accomodate the Ba^{2+} ion, and also to the presence of flexible CH_2 bridges between the rings [106].

In the case of polymer P|22|EpiDDP, which is insoluble in water, a liquid-liquid extraction process was used to determine the stability constants [110]. Also in this case each ligand unit maintains its independence, but according to the authors, this behaviour may be explained by the relatively low molecular weight of the polymer.

Studies on other polymers having polyester and polyamidic structures, and containing the same type of crown ethers, indicate that the polyamido structure leads to poor binding properties, while better properties on this respect are obtained with the polyesters [111]. The stability constants, in the case of polymers, depend on the nature of the comonomeric units, decreasing in the following order: adipic > sebacic > terephthalic. The triglycolic polyester appears to be a special case. Binding

Table 10. Basicity and stability constants of polymeric ligands and models in water at 20 °C in presence of 0.1 M NMe$_4$ Br

Cation	Ionic radii (Å)	Log K$_{s1}$ [22]EtOH	[21]EtOH	[222$_N$]EtOH	p[22]EpiO	p[22]EpiC	p[21]EpiC	p[222$_N$]EpiO
pk$_1$		8.44	8.64	9.99	8.76	8.33	8.27	9.13
pk$_2$		6.88	7.35	7.25	6.78	4.75	5.07	5.24
pk$_3$				3.30				2.00
pk$_4$				2.70				+ +
Na$^+$	0.98		1.0					0.9
K$^+$	1.33	1	1					1.4
Mg^{2+}	0.78	1	1.2					
Ca^{2+}	1.06	3.7	4.1		3.7	2.4	2.5	2.3
Sr^{2+}	1.27	4.3	3.8		4.2	2.9	2.5	3.4
Ba^{2+}	1.43	5.3	3.2	4.8	9.7[a]	3.6	2.5	4.1
Cd^{2+}	1.03	7.1	6.4	9.7	6.6	5.6	5.4	6
Cu^{2+}	0.72		8.1	13.5	7.3	6.7	6.5	11.3
Co^{2+}	0.92			4.9				
Ni^{2+}	0.78			4.8				
Zn^{2+}	0.80			5.6				

[a] K$'_{s2}$ calculated for 2:1 complex; Estimated cavity radii: [21]: 1.0 Å; [22]: 1.4 Å; [222$_N$]: 1.4 Å; K$_{s1}$ and K$'_{s2}$ are the stability constants of the 1:1 and 2:1 complexes, respectively

properties are better than those of other polyesters, and, more interestingly, the complexation is not selective. This can be understood if we consider a strong polymeric effect. The glycolic bridges participate in complex formation, and make it possible to accomodate cations of any size, thus leading to a poor selectivity.

4 Conclusions

A considerable amount of data on the protonation, and complex formation with metal ions of polymeric amines have been reported. A critical insight leads to the conclusion that much has to be done in order to reach a clear vision of the chemical properties of many polymers of this kind. Most protonation studies deal with the determination of basicity constants with potentiometric techniques, which alone give little information on the protonation mechanism; only few studies have been substantiated by spectroscopic (nmr) and calorimetric measurements.

On the whole, the same is true for most available studies on the complexing abilities of polymeric amines. As a consequence, in most cases it is not easy to predict "a priori" the possible applications of the amino polymers considered.

Acknowledgement: We thank Mr. Mauro Porcù for helpful technical assistance.

5 References

1. Reynolds, D. D., Kenyon, W. O.: J. Amer. Chem. Soc. *69*, 911 (1947); Katchalsky, A., Mazur, J., Spitnik, P.: J. Polym. Sci. *23*, 513 (1957)
2. Rath, H., Hilscher, E.: German Patent, 1, 158, 528 (1963)
3. Bayer, E., Geckeler, K., Weingartner, K.: Makromol. Chem.: *181*, 585 (1980)
4. Hart, R.: Makromol. Chem. *32*, 51 (1959); Bloys van Treslong, C. J., Morra, C. F. H.: Recl. Trav. Chim. Pays-Bas, *94*, 101 (1975)
5. Dawson, D. J., Gless, R. D., Wingard, R. E., Jr.: J. Amer. Chem. Soc. *19*, 5996 (1976)
6. Bolto, B. A., McNeill, R., Macpherson, A. S., Siudak, R., Weiss, D. E., Willis, D.: Austral. J. Chem. *21*, 2703 (1968)
7. Mark, H. S., Gaylord, N. G.: "Encyclopedia of Polymer Science and Technology" *1*, 734 (1964)
8. Dick, R. C., Ham, G. E.: J. Macromol. Sci. A4, 1301 (1970); Lukovkin, G. M., Pshezhetsky, V. S., Murtazaeva, G. A.: Europ. Polym. J. *9*, 559 (1973)
9. Saegusa, T., Ikeda, H., Fujii, H.: Polym. J. *3*, 35 (1972)
10. Lewis, E. A., Barker, J., St. Pierre, T.: Macromolecules, *14*, 546 (1980)
11. Bassiri, T. G., Levy, A., Litt, M.: Polym. Letters *5*, 871 (1967); Kagiya, T., Matsuda, M.: J. Macromol. Sci. Chem. *A5*, 1265 (1971)
12. Chatani, Y., Kobatake, T., Tadokoro, H., Tanaka, R.: Polym. Prepr. Japan *30*, 644 (1981)
13. Chatani, Y., Tadokoro, H., Saegusa, T., Ikeda, H.: Macromolecules *14*, 315, (1981)
14. Hagge, W. et al.: Belgium Patent 622, 716 (C.A. *34*, 3848)
15. Sheets, D. P.: U.S. Patent 3, 134, 740 (C.A. *61*, 7197 (1964))
16. Shepard, J., Kitchener, J. A.: J. Chem. Soc. 86 (1957)
17. Nonogaki, S., Makishima, S., Yoneda, Y.: J. Phys. Chem. *62*, 601 (1958)
18. Dingmann, J. Jr., Siggia, S., Barton, C., Hiscock, K. B.: Anal. Chem. *44*, 1351 (1972)
19. Jones, G. D., MacWilliams, D. C., Braxter, N. A.: J. Org. Chem. *30*, 1944 (1965)
20. Tanaka, R., Koike, M., Tsutsui, T., Tanaka, T.: J. Polym. Sci. Polym. Lett. Ed. *16*, 13 (1978)
21. Goethals, E. J., Schacht, E. H.: Polyamines and Polyammonium salts derived from azetidine monomers, in: Polymeric amines and ammonium salts (ed.) Goethals, E. J., p. 67, New York, Pergamon Press 1980, and references therein
22. Hagi, H., Ooishi, O., Tanaka, R.: Synthesis of linear poly(tertiary amine)s by selective dealkylation of poly(quaternary ammonium salt)s, in: Polymeric amines and ammonium salts (ed.) Goethals, E. J., p. 31, New York, Pergamon Press 1980, and references therein
23. Huguet, J., Vert, M., Spassky, N., Selegny, E.: Makromol. Chem. *170*, 23 (1973)
24. Huguet, J., Vert, M., Selegny, E.: Europ. Polym. J. *10*, 261 (1974)
25. Inove, S., Tsuruka, T., Furukawa, J.: Makromol. Chem. *53*, 215 (1962); Furukawa, J., Kawabata, N., Kato, A.: J. Polym. Sci.: B *5*, 1073 (1967)
26. Spassky, N., Sigwalt, P.: Tetrahedron Lett. 3541 (1968)
27. Tsunetsugu, T., Furukawa, J., Fueno, T.: J. Polym. Sci. A1, *9*, 3541 (1971)
28. Vallin, D., Huguet, J., Vert, M.: 26th IUPAC Symp. on Macromolecules, Mainz/FRG (1979) preprint n °C3 16
29. Huguet, J., Vert, M.: J. Polym. Sci. *14*, 1257 (1976)
30. see for instance Natta, G., Mazzanti, G., Longi, P., Dall'Asta, G., Bernardini, F.: J. Polym. Sci. *51*, 487 (1961); Natta, G., Mazzanti, G., Dall'Asta, G., Longi, P.: Makrom. Chemie *37*, 160 (1960)
31. Strauss, U. P., Assony, S. A., Jackson, E. G., Leyton, L. H.: J. Polym. Sci. *9*, 509 (1952); Fuoss, R. M., Watanable, M., Coleman, B. O.: J. Polym. Sci. *48*, 5 (1960)
32. Sugii, A., Ogawa, N., Iinuma, Y., Yamamura, H.: Talanta *28*, 551 (1981)
33. Nishide, H., Tsuchida, E.: Makromol. Chem. *177*, 2295 (1976)
34. Pasini, C., Vercellone, A.: Gazz. Chim. Ital. *85*, 349 (1955)

35. Overberger, C. G., Vorchheimer, N.: J. Amer. Chem. Soc. 85, 951 (1963)
36. Gregor, H. P., Gold, D.: J. Amer. Chem. Soc. 61, 1347 (1957)
37. Tsuchia, E., Nishide, H.: Selective adsorption of metal ions to polymeric amines immobilized by template reaction, in: Polymeric amines and ammonium salts (ed.) Goethals, E. J., p. 271, New York. Pergamon Press 1980
38. Danusso, F., Ferruti, P.: Polymer 11, 88 (1970)
39. Danusso, F., Ferruti, P., Ferroni, G.: Chimica Industria (Milan) 49, 271 (1967)
40. Danusso, F., Ferruti, P., Ferroni, G.: Chimica Industria (Milan) 49, 453 (1967)
41. Danusso, F., Ferruti, P., Ferroni, G.: Chimica Industria (Milan) 49, 587 (1967)
42. Barbucci, R., Ferruti, P., Improta, C., Delfini, M., Segre, A. L., Conti, F.: Polymer 19, 1329 (1978)
43. Barbucci, R., Ferruti, P., Improta, C., La Torraca, M., Oliva, L., Tanzi, M. C.: Polymer 20, 1298 (1979)
44. Barbucci, R., Ferruti, P.: Polymer 20, 1061 (1979)
45. Danusso, F., Ferruti, P., Ferroni, G.: Chimica Industria (Milan) 49, 826 (1967)
46. Ferruti, P., Brzozowski, Z.: Chimica Industria (Milan) 50, 441 (1968)
47. see for instance, Barbucci, R. et al., Macro-inorganics: coordination compounds with poly-(amido-amine)s, in: Polymeric Amines and Ammonium Salts, (ed.) Goethals, E. J., p. 263, New York, Pergamon Press 1980
48. Gramain, P., Frere, Y.: Polymer 21, 921 (1980)
49. Gramain, P., Frere, Y.: Macromolecules 12, 1038 (1979)
50. Gramain, P., Frere, Y.: Makromol. Chem. Rapid Commun. 2, 161 (1981)
51. Gramain, P., Kleiber, M., Frere, Y.: Polymer 21, 915 (1980)
52. Blasius, E., Janzen, K. P., Lukenburger, H., Nguyen, V. B., Klotz, H., Stockemer, J.: Journal of Chromatography 167, 307 (1978)
53. see for instance, Card, R. J., Neckers, D. C.: J. Amer. Chem. Soc. 99, 7733 (1977); Valera, N. S., Hendricker, D. G.: Polymer 22, 1007 (1981); Drago, R. S., Gaul, J., Zombeck, A., Straub, D. K.: J. Amer. Chem. Soc. 30, 1033 (1980)
54. Bloys van Treslong, C. J.: J. Recl. Trav. Chim. Pays-Bas 97, 13 (1978)
55. Bloys van Treslong, C. J., Staverman, A. J.: J. Recl. Trav. Chim. Pays-Bas 93, 171 (1974)
56. Katchalsky, A., Rosenbeck, K., Altmann, B.: J. Pol. Sci. 23, 955 (1957)
57. Van den Berg, J. W. A., Bloys van Treslong, C. J., Polderman, A.: J. Recl. Trav. Chim. Pays-Bas 92, 3 (1973)
58. Rinaldi, P. L., Yu, C., Levy, G. C.: Macromolecules 14, 551 (1981)
59. Barone, V., Barbucci, R., Russo, N.: Gazz. Chim. Ital. 111, 115 (1981)
60. Bayer, E., Geckeler, K., Weingärtner, K.: Makromol. Chem. 181, 585 (1980)
61. Teyssié, Ph., Decoene, C., Teyssié, M. T.: Makromol. Chem. 84, 51 (1965)
62. Bolto, B. A., Mc Neill, R., Macpherson, A. S., Siudak, R., Weiss, D. E., Willis, D.: Aust. J. Chem. 21, 2703 (1968)
63. Dingman, J., Siggio, S., Barton, C., Hiscock, K. B.: Anal. Chem. 44, 1351 (1972); and references therein
64. Huguet, J.: Nov. J. de Chimie 3, 293 (1979)
65. Shatkay, A., Michaeli, I.: J. Phys. Chem. 70, 3777 (1966)
66. Huguet, J., Vert, M.: Europ. Polym. J. 12, 469 (1976)
67. Blout, E. R., Bovey, F. A., Goodman, M., Lotan, N.: "Peptides, Polypeptides and Proteins" J. Wiley & Sons, N. Y. (1974) p. 166–176
68. Huguet, J., Vert, M.: Europ. Polym. J. 12, 469 (1976)
69. Vallin, D., Huguet, J., Vert, M.: Acid-base and chiroptical properties of partially quaternized optically active poly[thio-1-(N,N-diethyl-aminomethyl)ethylene], in: Polymeric amines and ammonium salts, (ed.) Goethals, E. J., p. 219, New York, Pergamon Press 1980
70. Vallin, D., Huguet, J., Vert, M.: Polym. J. 12, 113 (1980)
71. Engel, J., Schwarz, G.: Angew. Chem. Internat. Ed. 9, 389 (1970)
72. Kirsh, Y. E., Komarova, O. P., Lukovkin, G. M.: Europ. Polym. J. 9, 1405 (1973)
73. Kirkwood, J. G., Westheimer, F. M.: J. Chem. Phys. 6, 506 (1938)
74. Puterman, M., Koenig, J. L., Lando, J. B.: J. Macromol. Sci.-Phys. B16, 89 (1979)
75. Muller, G., Ripoll, C., Sélégny, E.: Europ. Polym. J.: 7, 1373 (1971)
76. Muller, G.: "Polyelectrolytes" Ed. E. Sélégny, D. Reidel Publ. Co. Dordrecht-Holland (1974) p. 195–205

77. Puterman, M., Garcia, E., Lando, J. B.: J. Macromol. Sci. Phys. *B16* (1) 117 (1979)
78. Pino, P., Ciardelli, F., Lorenzi, G. P., Natta, G.: J. Amer. Chem. Soc. *84*, 1487 (1962)
79. Helfferrich, F.: "Ion Exchange" McGraw-Hill (1962)
80. Metayer, M., Chabot, F., Sélégny, E.: J. Chim. Phys. *76*, 404 (1979)
81. Kirsh, E., Pavlova, N. R., Kabanov, V. A. Europ. Polym. J.: *11*, 47 (1975)
82. Joyce, D. E., Kurucsev, P.: Polymer *21*, 1451 (1980)
83. Nishikawa, H., Tsuchida, E.: J. Phys. Chem. *79*, 2072 (1975)
84. Tsuchida, E., Nishide, H., Nishiyama, T.: J. Polymer Sci., Sympos. *47*, 35 (1974)
85. Takagushi, T., Klotz, I. M.: Biopolymers *11*, 483 (1972)
86. Nishide, H., Tsuchida, E.: Makromol. Chem. *177*, 2295 (1976)
87. Joyce, D. E., Kurucsev, T.: Polymer *21*, 1451 (1980)
88. Bolto, B. A.: The role of polyamine resins in thermally regenerable ion exchange, in: Polymeric amines and ammonium salts, (ed.) Goethals, E. J., p. 365, New York, Pergamon Press 1980
89. Gregor, H. P., Gold, D. H.: J. Phys. Chem. *61*, 1347 (1957); Gold, D. H., Gregor, H. P.: J. Phys. Chem. *64*, 1464 (1960); ibidem J. Phys. Chem. *64*, 1461 (1960)
90. Liu, K., Gregor, H. P.: J. Phys. Chem. *69*, 1252 (1965)
91. Bjerrum, J.: "Metal Amine Formation in Aqueous Solutions" P. Haas and Son, Copenhagen, 1941
92. Gregor, H. P., Luttinger, L. B., Loebl, E. M.: J. Phys. Chem. *59*, 34 (1955)
93. Sato, M., Kondo, K., Takemoto, K.: Makromol. Chem. *179*, 601 (1978)
94. Sato, M., Kondo, K., Takemoto, K.: J. Macromol. Sci. Chem. *A14* (3), 349 (1980)
95. Sato, M., Shindo, H., Kondo, K., Takemoto, K.: J. Polym. Sci., Pol. Chem. Ed. *18*, 101 (1980)
96. Barbucci, R., Barone, V., Ferruti, P.: Atti Accad. Naz. Lincei, Rend. Classe Sci. Fis. Mat. Nat. *64*, 481 (1978)
97. Barbucci, R., Casolaro, M., Ferruti, P., Barone, V., Lelj, F., Oliva, L.: Macromolecules *14*, 1203 (1981)
98. Barbucci, R., Ferruti, P., Micheloni, M., Delfini, M., Segre, A. L., Conti, F.: Polymer *21*, 81 (1980)
99. Barbucci, R., Barone, V., Ferruti, P., Oliva, L.: J. Polymer Sci. Symposia *69*, 49 (1981)
100. Ferruti, P., Oliva, L., Barbucci, R., Tanzi, M. C.: Inorg. Chim. Acta *41*, 25 (1980)
101. Barbucci, R., Barone, V., Ferruti, P., Delfini, M.: J. C. S. Dalton 253 (1980)
102. Barbucci, R., Casolaro, M., Ferruti, P., Barone, V.,: Polymer *23*, 148 (1982)
103. Ferruti, P., Riolo, C., Soldi, T., Pesavento, M., Barbucci, R., Beni, M. C., Casolaro, M.: J. Appl. Poly. Sci. in the press
104. Barbucci, R., Casolaro, M., Beni, M. C., Ferruti, P., Pesavento, M., Soldi, F., Riolo, C.: J. C. S. Dalton 2559 (1981)
105. Dietrich, B., Lehn, J. M., Sauvage, J. P.: Tetrahedron *29*, 1647 (1973)
106. Gramain, Ph., Frere, Y.: Polymer *21*, 921 (1980)
107. Gramain, Ph., Frere, Y.: Ind. Eng. Chem. Prod. Res. Dev. in the press
108. Lehn, J. M.: Struct. Bonding (Berlin) *16*, 1 (1973)
109. Lehn, J. M., Montavon, F.: Helv. Chim. Acta *61*, 67 (1978)
110. Gramain, Ph., Frere, Y.: Nouv. J. Chimie *3*, 59 (1979)
111. Gramain, Ph., Kleiber, M., Frere, Y.: Polymer *21*, 915 (1980)

Received October 13, 1982
G. Dall'Asta (editor)

Polymerization of Quinodimethane Compounds

Shouji Iwatsuki
Dept. of Chemical Research for Resources, Faculty of Engineering, Mie University,
Kamihama-cho, Tsu, 514, Japan

This article surveys the polymerization of quinodimethanes including flash pyrolysis of p-xylene, the synthesis of poly-p-xylylene other than by flash pyrolysis, the vapor-coating process by Gorham, and the polymerizations of halo-p-xylylenes and electron-accepting quinodimethanes.

Advances in Polymer Science 58
© Springer-Verlag Berlin Heidelberg 1984

1 Introduction

Since Szwarc successfully prepared poly-p-xylylene in 1947 by flash pyrolysis of p-xylene, p-xylylenes[1] and their polymers have become increasingly important due to their unusual chemical and physical properties. Errede and Szwarc [i] reviewed the polymerization of p-xylylene in 1957. After the publication of their review article, various progress in this field has been made, e.g. the vapor-coating technique by Gorham, and the polymerization behaviors of halo-p-xylylenes and electron-accepting quinodimethanes. Therefore, I review again the polymerization of quinodimethanes taking into account recent developments.

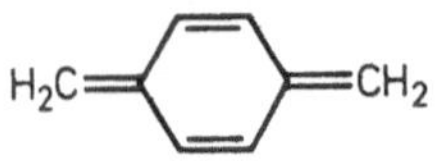

$$H_2C=\!\!\!\!\left\langle\!\!\!\!\bigcirc\!\!\!\!\right\rangle\!\!\!\!=\!CH_2$$

p-Xylylene (QM)

2 Flash Pyrolysis of p-Xylene

In 1947 Szwarc prepared a white polymeric material [1] by rapid flow pyrolysis of p-xylene under reduced pressure. On the basis of p-xylylene diiodide [2] detected in the reaction mixture of the pyrolysis products with iodine gas he proposed a formation [1,3] of p-xylylene(p-quinodimethane) (QM) in this pyrolysis. He claimed the polymeric material to be poly-p-xylylene(poly-QM) [1] and proposed a mechanism [2] for the formation of poly-QM, involving thermal cleavage of carbon-hydrogen bonds of p-xylene to yield p-xylyl radicals which collide with each other to give p-xylene and QM through disproportionation. QM condenses and polymerizes to produce poly-QM.

The pyrolysis of p-xylene even at the extremely high temperature of 1150 °C was found to give the polymeric product only in a low yield (25%) [4]. If the pyrolysis is carried out at 900 °C, 85% of p-xylene remain unchanged and only 8.8% of it are converted into poly-QM. Several compounds such as benzene, toluene, di-xylylene, di-p-xylyl, and low molecular weight resin are formed as by-product, [4] total amount of which goes up to 10–20% of the polymeric material poly-QM. The obtained poly-QM has a high melting point and is inert to organic and inorganic reagents. This material has been extensively studied [4-9]. It has also attracted the interest of many quantum chemists [10-13]. Coulson [10] calculated the energy difference of QM molecule between the singlet ground state and the triplet excited state to be 8–9 kcal. The corresponding value for ethylene was determined by Evans as 82 kcal [14]. This unusually low energy difference is responsible for the very high reactivity of the QM molecule.

---

1 The compound which is customarily called p-xylylene or p-quinodimethane is denoted as 1,4-dimethylene-2,5-cyclohexadiene under the IUPAC nomenclature system. However, since in this review article reference is frequently made to the orginal papers, the trivial names are retained.

In a series of studies [15-25] on QM compounds, Errede reported a method [15] for the preparation of a relatively stable solution of QM, which involves rapid flow pyrolysis of p-xylene at ca. 1000 °C under low pressure (4 mmHg) and subsequent instantaneous condensation of the pyrolysis products into a solvent maintained at −78 °C. In addition to QM, the following pyrolysis products were detected [17]; toluene, styrene, p-ethylstyrene, 1,2-(di-p-tolyl)ethane, diacrylmethane, anthracene, and 4,4′-dimethylstilbene. When oxygen or air is bubbled through a solution of QM, poly-QM peroxide is formed with an oxygen content ranging from 1 to 23% (the corresponding molar ratios of QM to oxygen were calculated to be 31 to 1 up to 1 to 1) [16]. QM was found to be copolymerizable with pseudomonomers such as sulfur dioxide [17], nitroso compounds [17], and phosphorous trichloride [23]. However, since QM is much more reactive than conventional olefinic monomers, provided both monomers are mixed in the usual way, copolymerization does not occurs at −78 °C but the homopolymer of QM is obtained [17]. In such unusual way that solution of QM kept at −78 °C is added to a solution of conventional monomer maintained at about 100 °C a copolymer can be formed [17]. Conventional chain transfer agents such as carbon tetrachloride, chloroform, p-cumene, nitrobenzene, and hydroquinone are not sufficiently active in the polymerization of QM [17]. When a threefold excess of thiophenol is added to QM, chain transfer reaction occurs yielding a telomer with a ratio of QM to thiophenol units of 21:1 [17]. The polymerization of QM proceeds slowly even at temperatures as low as −78 °C [20]. When the monomer

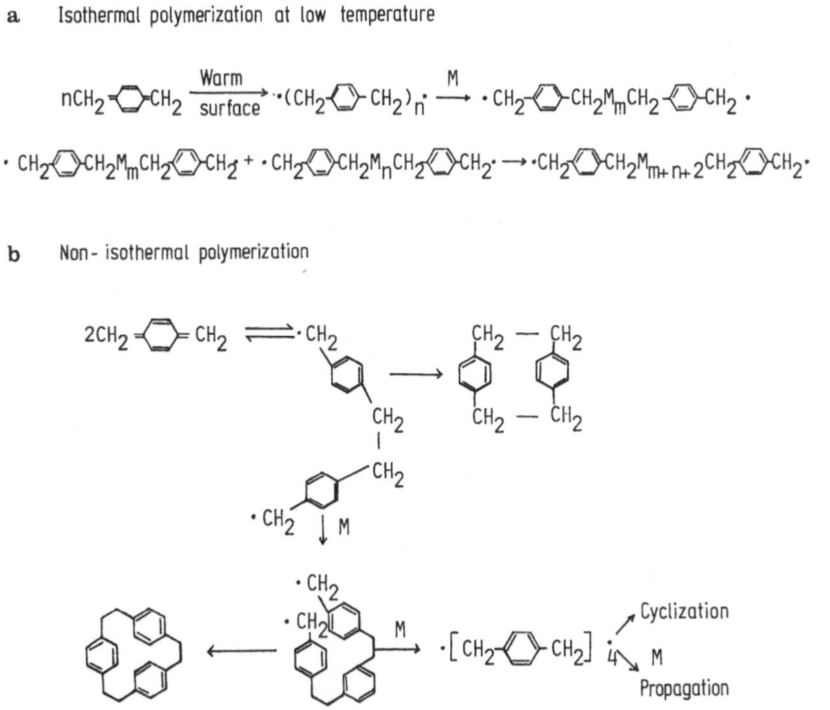

a   Isothermal polymerization at low temperature

$$nCH_2{=}\langle{=}\rangle{-}CH_2 \xrightarrow[\text{surface}]{\text{Warm}} \cdot\cdot(CH_2{-}\langle{=}\rangle{-}CH_2)_n\cdot \xrightarrow{M} \cdot CH_2{-}\langle{=}\rangle{-}CH_2{}_m CH_2{-}\langle{=}\rangle{-}CH_2\cdot$$

$$\cdot CH_2{-}\langle{=}\rangle{-}CH_2{}_m CH_2{-}\langle{=}\rangle{-}CH_2\cdot + \cdot CH_2{-}\langle{=}\rangle{-}CH_2{}_n CH_2{-}\langle{=}\rangle{-}CH_2\cdot \rightarrow \cdot CH_2{-}\langle{=}\rangle{-}CH_2{}_{m+n}+ 2 CH_2{-}\langle{=}\rangle{-}CH_2\cdot$$

b   Non-isothermal polymerization

Fig. 1a and b. Schematic representation of the polymerization of QM solution is warmed to room temperature, the polymerization occurs extremely rapidly. Polymer molecules with a degree of polymerization higher than about 20 precipitate from the solution. Besides, when a cold solution of QM is brought in contact with a warm surface for brief moment, polymerization of QM is initiated at −78 °C, presumably involving formation of a diradical with n-mers which continues to grow by successive addition of QM monomer and/or by coupling. Propagation continues at −78 °C until all monomer is consumed or the free radical end groups are entrapped in the chain mesh, giving a linear polymer with molecular weight greater than 2×10^5 (calculated from the number of radioactive iodine end groups incorporated into the polymer when the reaction mixture is quenched with labeled iodine after polymerization has proceeded to 95 % completion) [20]. The polymerization was found to be first order with respect to monomer and second order with respect to free radical end groups [20]. When QM is subjected to polymerization in solution at temperatures higher than −78 °C, some soluble low molecular weight products such as cyclo-tri-QM, cyclo-tetra-QM, 1,4-bis(2-p-tolylethyl)benzene, and low molecular weight polymer are formed in addition to insoluble high molecular weight polymer. When a cold solution of QM is added dropwise to an inert solvent such as toluene kept at about 100 °C, cyclo-di-QM is obtained in good yield [20]. Therefore, the proper control of the reaction conditions allows a selective formation of low

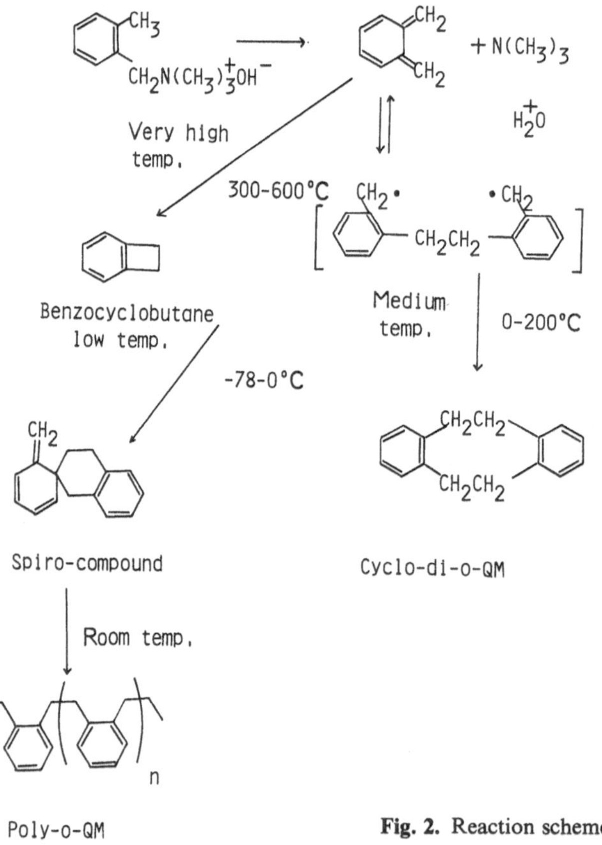

Fig. 2. Reaction scheme of the polymerization of 0-QM

molecular weight cyclic products such as cyclo-tri-QM and cyclo-di-QM. The proposed reaction scheme is shown in Fig. 1. A kinetic study of the polymerization at low temperature yields an apparent rate constant (kP) of $9 \pm 1 \times 10^{-6}$ s^{-1} according to the equation dM/dt = —kPM where k is the rate constant for the addition of monomer to radical end group, M the concentration of monomeric QM, and P the number of free radical end groups. The plot of the log kP vs. the reciprocal of the absolute temperature gives a straight line, from the slope of which an activation energy of polymerization is calculated as 8.7 kcal [20]. Termination which decreases in number of polymerization sites is not observed over a period of 10 hr, but it becomes appreciable when the polymerization is continued for a longer period. The apparent rate constant (kP) decreases slowly but steadily. When the reciprocal of the apparent rate constant is plotted against time of polymerization, a straight line is obtained, indicating that the disappearance of free radical end groups is second order. In the kinetic analysis, the ratio of rate constant of radical coupling (termination) to that of radical addition to monomer (propagation) is calculated to be 0.45 [20], which implies that propagation and termination are similar type of reactions with rate constants of similar figures. This is in sharp contrast to the conventional free-radical vinyl polymerization.

Furthermore, Errede successfully prepared o-xylylene (o-quinodimethane) (o-QM) by Hofmann degradation of o-methylbenzyl-trimethylammonium hydroxide at low pressure using a modified flow process [22]. When o-QM is warmed from —78 °C to room temperature, spiro-(5,6)-2,3-benz-6-methyleneundeca-7,9-diene(spiro-di-QM) is obtained in a good yield (61 %), together with cyclo-di-o-QM, bis-(o-methylbenzyl) ether, a mixture of esters (similar to benzyl benzoates) and o-methylbenzyl alcohol. When o-QM is heated at temperatures from 0 to 200 °C, cyclo-di-o-QM is obtained. Furthermore, at 300 to 600 °C, benzocyclobutane is predominantly formed. Apparently, the formation of the spiro compound is favored at temperatures lower than 0 °C. A spiro-dimer can be preserved without any change at —15 °C. It solidifies slowly when cooled to —20 °C and remelts at —5 to 0 °C. When it is warmed to room temperature, it polymerizes very slowly to give high molecular weight poly-o-QM with the intrinsic solution viscosity of about 0.6 and glass temperature of 9 °C [22]. The spiro-dimer is copolymerized with conventional olefinic monomer such as styrene, acrylonitrile, methyl methacrylate, vinylidene fluoride, and butadiene [22]. In the polymerization of the spiro-dimer, effective chain transfer with conventional reagents such as mercaptans and carbon tetrahalides [22] takes place in contrast to QM polymerization. When an acid catalyst is added to a concentrated solution of the spiro-dimer in hexane, poly-o-QM is obtained, whereas in the case of a diluted solution of the spiro-dimer, 1-methyl-dibenzo(a,d)cyclohepta-1,4-diene is producced in good yield via intramolecular aromatic substitution of o-(β-tolylethyl)-benzyl carbonium ion, formed by rearrangement of the proton addition product of the spiro-dimer [25]. Spiro-dimer is copolymerized with formaldehyde in the presence of an acid catalyst to give the corresponding polymeric ether [25].

3 Preparations of Poly-p-Xylylenes other than by Pyrolysis

Amorphous, low molecular weight ($\sim$3000) poly-p-xylylene (poly-QM) was synthesized by the Wurts-Fittig reaction of 7,8-dibromo-p-xylene with sodium or magnesium metal [26]. This reaction was improved in various ways to obtain satisfactory yield of the crystalline polymer [9,27]. Thus, different dehalogenation agents were used, e.g. reduced iron and cobalt powder suspended in water [28], Urushibara Nickel [28], naphthalenealkali complex [29], and tin(II) chloride [30]. Electrolytic reduction also was applied [31]. When strong bases such as sodium amide [32,33], potassium tert-but-oxide [34], and sodium methoxide [35] are used, 7,8-dihalo-p-xylene is converted to the copolymer of xylylidene and halo-QM. This copolymer has the chemical formula,

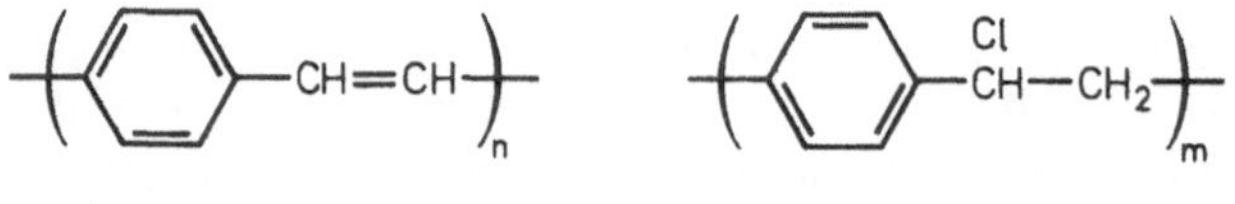

The reaction of 7-chloro-p-xylene with potassium tert-butoxide in p-xylene in the presence of stable N-oxy biradicals such as 2,2,6,6-tetramethylpiperidinoxy-4-spiro-2'-(1',3'-dioxane)-5'-spiro-5''-(1'',3''-dioxane)-2''-spiro-4''-(2''',2''',6''',6'''-tetramethyl piperidinoxy) gives the corresponding copolymer [36] indicating the formation of QM as a reaction intermediate.

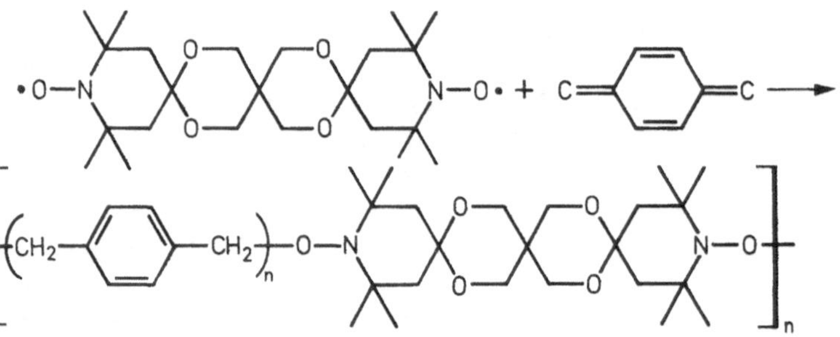

The Friedel-Craft reaction of benzene with 1,2-dichloroethane affords a polymer which is insoluble in any solvent and does not melt, i.e. no linear polymer has been formed [37,38]. The same reaction of 2-chloroethylbenzene likewise yields only an intractable polymer [38]. However, the Hofmann degradation of p-methy-benzyltri-methylammonium hydroxide at 60 ~ 100 °C gives linear and soluble poly-QM in a high yield [39]. This process may widely be applied to the synthesis of other QM polymers, e.g. poly-2,5-dimethoxy-QM, which may be hydrolyzed to poly-2,5-dihydroxy-QM. This polymer exhibits redox properties [40]. In a modified Hofmann degradation, [p-[(trimethylsilyl)methyl]benzyl] trimethyl ammonium ioide can decompose with tetrabutylammonium fluoride in acetonitril at refluxing temperature and at room temperature to give cyclo-di-QM (50 % yield) and poly-QM (51 % yield) with cyclo-di-QM (6 % yield), respectively [41]. Di-azo-p-tolylmethane may undergo a cationic rearrangement to poly-QM [42].

4 Pyrolysis of Paracyclophane: The Vapor-Coating Process

Gorham [43-45] synthesized poly-p-xylylene(poly-QM) by a new technique involving vacuum pyrolysis of [2,2]-paracyclophane, cyclo-di-xylene (cyclo-di-QM) at 600 °C. When the pyrolyzed gas of cyclo-di-QM, namely QM, is condensed on glass or metal surface, the polymerization of the generated QM yields a tough, transparent polymeric film. Thus, the process is called vapor-coating process. Compared with the Szwarc-Errede direct pyrolysis, this process offers the following advantages: It affords two molecules of QM in quantitative yield, produces linear polymers free of cross-linking and low molecular weight by-products and the obtained polymers are soluble in hot chlorinated biphenyls. Due to the considerably lower pyrolysis temperature, the vapor-coating process may be applied to synthesis of a variety of substituted QM polymers.

Cyclo-di-QM was indentified first by Brown and Farthing [9,46,47]. In the Szwarc's pyrolysis of p-xylene under reduced pressure at 680 to 850 °C, the products obtained in 10–20% yield were mainly polymeric material which was found to contain a small portion of low molecular weight compounds soluble in chloroform. Thus, 4,4-dimethyldibenzyl was detected in the chloroform extract [7]. In addition, traces of compounds insoluble in acetone were found [7], whose structure was determined as [2,2]-paracyclophane by X-ray diffrection measurements [47]. Independently, Cram and Steinberg prepared a cyclic dimer in a poor yield by Wurtz-Fittig reaction of 4,4'-dibromomethyl dibenzyl [48]. Due to its distorted structure this dimer is strically hindered and generally referred to as cyclophane. It has widely been studied [49,50]. Errede et. al. prepared selectively cyclo-di-QM by adding dropwise a 0.1 M QM solution in hexane maintained at −78 °C to toluene heated at 100 °C [20]. Pollart developed a solvent quenching technique for the synthesis of cyclo-di-QM [51,52]. The condensation of QM vapor directly into an organic solvent at a temperature of 50 to 200 °C results the formation of cyclo-di-QM in a yield higher than 90% [51]. The rapid pyrolysis of a mixture of steam and p-xylene at 850–900 °C followed by condensation of the vapor in an organic solvent such as p-xylene at 50 °C gives cyclo-di-QM in 8–10% yield with only 0.1% polymeric material [52].

Gorham prepared about 30 kinds of substituted paracyclophanes including dichloro, dibromo, dicyano, dimethyl, diethyl, and tetrachloro derivatives [43].

The various substituted QM monomers condense and polymerize on the surface at temperatures lower than each specific temperature referred to as threshold condensation temperature which is related to the molecular weight and volatility of the respective monomer. The threshold condensation temperature is defined as the highest temperature of the surface on which the QM monomers condense and polymerize at an appreciable rate. At normal pressure (about 0.1 mmHg) of the system the threshold temperatures are determined as follows; 30 °C for QM, 60 °C for 2-methyl-QM, 90 °C for 2-ethyl- and 2-chloro-QM and 130 °C for 2-cyano-, 2-bromo-, and dichloro-QM.

The pyrolysis of unsymmetrically substituted cyclo-di-QM and the subsequent polymerization of the generated monomers by means of the vapor-coating process has been studied [43]. The pyrolysis gas of acetyl-cyclo-di-QM is initially led through a glass tube maintained at 90 °C and subsequently through another glass tube kept at a temperature of 20 °C. The obtained polymer at 90 °C has been identified as poly-acetyl-QM on the basis of elemental analysis and IR spectrometry. A second pyrolysis process and the subsequent fractional polymerization at the threshold polymer obtained at 25 °C has been characterized as poly-QM by IR spectrometry. It melts at 400 °C and is insoluble in all organic solvents below 250 °C. These results have been explained by the formation of two species, QM and acetyl-QM, in the

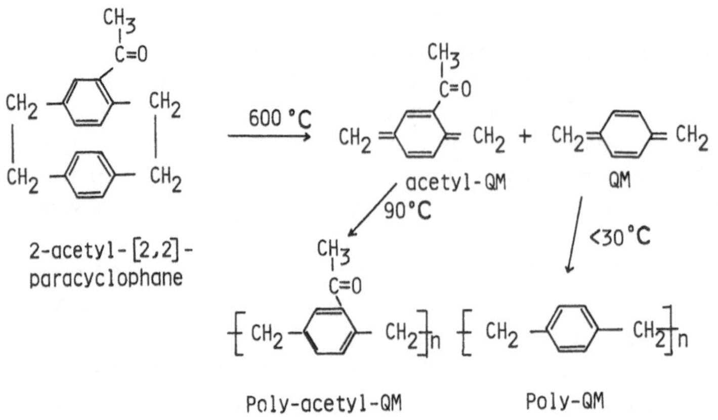

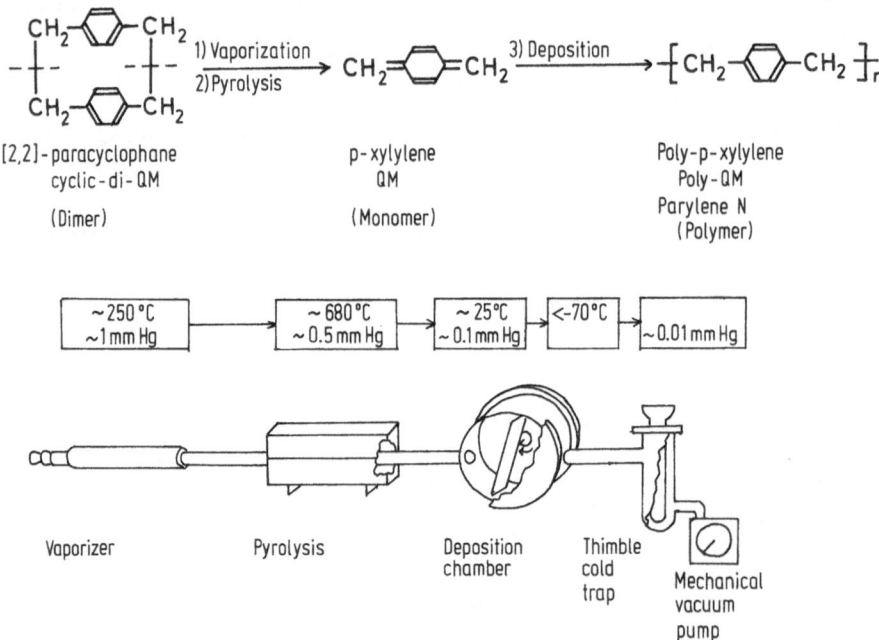

Fig. 3. Reaction scheme of the pyrolysis of 2-acetyl-[2,2]-paracyclophane and its fractional polymerization

Fig. 4. Schematic representation of the vapor-coating process, using [2,2]-paracyclophane as an example

condensation temperatures. The polymer synthesized by the vapor-coating process was found to be paramagnetic (radical concentration of 5 to 10×10^{-4} mol per mol QM). When the polymer is heated at elevated temperature, the ESR signal disappears. Therefore, this polymerization has been suggested to proceed by a free radical mechanism similarly as described by the scheme of Errede [20].

Poly-QM obtained by the use of the vapor-coating process is readily soluble in chlorinated biphenyls and benzyl benzonate at 300 °C and highly crystalline. The

Table 1. Physical and electrical properties of poly-QMs

	Poly-p-xylylene (Parylene N) [43]	Polychloro-p-xylyiene (Parylene C) [43]	Poly 7,7,8,8-tetrafluoro-p-xylylene [54]
Tensile properties (at room temperature)			
Tensile strength (psi)	6,800	10,600	6,200
Tensile modulus (psi)	350,000	460,000	360,000
elongation at break (psi)	10–15	220	100
Tensile modulus at 200 °C (psi)	25,000	25,000	—
Thermal properties			
crystalline melting point (°C)	400	290	
glass transition temperature (°C)	80	80	90
Permeability at 77 °F cm^3 (STP) mil/100 in^2 24 hr			
H_2	250	200	
CO_2	225	21	
O_2	30	8	
N_2	9	1	
H_2O (g mil/100 in^2 24 hr)	6.0	0.6	
Electrical Properties (1–3 mil film)			
Dielectric constant (1 Kc/s)	2.65	3.2	2.36
Dissipation factor (1 Kc/s)	0.0002	0.04	0.0008
Dielectric Strength (v/mil)	7,000	5,000	5.250

polymer film deposited at room temperature has been found to exist in metastable α-modification which is transformed to a stable β-modification upon heating to 220 °C or higher [53]. Single crystals with both modifications have been prepared from a 0.05 wt/vol% chloronaphthalene solution. The α-type crystals are composed of orthorhombic cells with a = 21.3 Å, b = 33.6 Å, and c = 6.58 Å, the β-type crystals are hexagonal with a = 20.52 ± 0.05 Å, and c = 6.58 ± 0.020 Å [53].

The vapor-coating process was developed by Union Carbide Corporation which commercially manufactures the unusual polymers under the trade name "Parylene"; Parylene N refers to unsubstituted poly-QM and Parylene C to poly-2-chloro-QM. The process involves three steps as outlined in Fig. 4 for Parylene N. The film deposited on the surface in this process is free of pinhole and can be ajusted to a thickness of several submicrons to several millimeters. The physical and electrical properties of these polymers are compiled in Table 1. They exhibit high moduli at room temperature and tensile moduli above 300,000 psi. Their glass transition temperatures are in the range of 60 to 90 °C and their melting points are as high as 290–400 °C. These polymers have low gas-permeability characteristics especially for Parylene C. Parlylene N is a dielectric exhibiting a very low dissipation factor, a high dielectric strength and a dielectric constant invariable with frequency. It is used as a dielectric of a plastic-film capacitor. Parylene C additionally exhibits a very low permeability to moisture and other corrosive gases and is assumed to be useful for the coating of critical electric assemblies.

7,7,8,8-Tetrafluoro-QM polymer was prepared by a similar process starting from the corresponding paracyclophane derivatives [54,55]. This polymer also exhibits physical and electrical properties similar to those of Parylene polymers (in Table 1). It is extremely resistant to sun light even after exposure for 3600 hr while Parylene N is changed to a brittle material after exposure for 535 hr [54].

5 Polymerization of Halo-p-Xylylenes

7,7,7,8,8,8-Hexachloro-p-xylene is dechlorinated on a copper mesh under reduced pressure (0.1 to 1.0 mmHg) at 300 to 600 °C to 7,7,8,8-tétrachloro-p-xylylene (TCX) in yields up to 90% [56]. The pyrolysis product is absorbed in toluene maintained

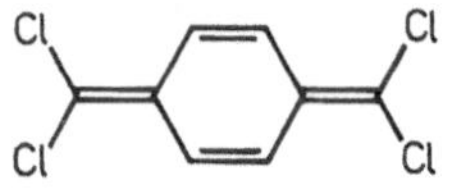

7,7,8,8 – Tetrachloro-p-xylylene (TCX)

at —78 °C. The resulting yellow colored suspension is cooled down and the obtained yellow precipitate is isolated by filtration. Repeated recrystallization from tetrahydrofuran under nitrogen at temperatures ranging from —60 to 0 °C gives yellow needles. When these crystals are kept at room temperature, their yellow color gradually faded and poly-TCX is formed. A tetrahydrofuran solution of 0.0244 M TCX polymerizes at 20 °C for 30 min up to a conversion of 50% [56]. The gaseous pyrolysis product is condensed on the surface maintained at a temperature above 120 °C to give a transparent film of poly-TCX [57]. The freshly prepared film still contains monomeric TCX and exhibits a strong ESR signal, indicating the presence of free radicals. Polymerization is completed by annealing of the film at 190 °C for 30 min [57]. When the pyrolysis gas is deposited on the surface below 90 °C, crystalline monomeric TCX is formed which gradually polymerizes [57]. At 100 °C, crystals and a large amount of a transparent film are obtained simultaneously [57]. The poly-TCX displays the following mechanical and electrical properties [57]; tensile modulus 480,000 psi, tensile strength 8,000 psi, softing range of 280–290 °C, and dielectric constant 2.81, and dissipation factor 2.6×10^{-4} at a frequency range of 60 cycles to 100 kilocycles. TCX may be recrystallized and kept at temperatures below —10 °C without polymerization. It is therefore clearly more stable than QM but much more reactive than conventional olefinic monomers.

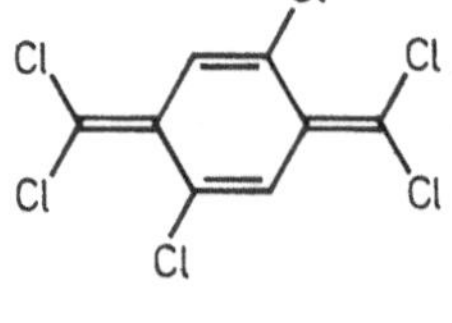

2,5,7,7,8,8 – Hexachloro-p-xylylene (HCX)

2,5,7,7,8,8-Hexachloro-p-xylylene(HCX) is prepared by passing 2,5,7,7,7,8,8,8-octachloro-p-xylene vapor over a copper mesh at 500 °C at a reduced pressure [58]. Pure yellow crystalline HCX is obtained by the same recrystallization method used for the isolation of TCX. HCX can be kept without any change below 0 °C for a long time. At room temperature, it changes gradually to a white powder which has been identified as poly-HCX on the basis of its elemental analysis. The change of HCX crystals to its amorphous polymer on standing was examined by X-ray diffraction measurement in order to follow the rate of the solid state polymerization of HCX. The height (or intensity) of the X-ray diffraction profile, corresponding to the crystalline portion (monomer) in the probe, decreased linearly with time, indicating the zero reaction order of the spontaneous solid state polymerization. After 66 hr the X-ray diffraction disappeared completely. This reaction order can be explained by the effective monomer concentration in the solid state. Thus, when an active site migrates into a crystal in the course of polymerization, the number of HCX monomers around the active site, which are susceptible to polymerization, is considered to be constant and a polymer chain already formed does not influence this number because it always exists just in rear of the active site and is excluded from the HCX crystal. In addition, the formation of the active sites is regarded to be independent of the number of HCX monomers in the volume unit of the crystal. The rate of spontaneous polymerization of HCX in benzene follows the first order kinetics with respect to monomer concentration. The apparent first order rate constants are 5.56×10^{-5} s^{-1} at 30 °C, 13.33×10^{-5} s^{-1} at 50 °C, etc. The Arrhenius plot of these rate constants gives a straight line from the its slope of which the apparent activation energy of the polymerization of HCX is calculated as 8.23 kcal/mol. The temperature at which the apparent first order constant of the polymerization of QM in toluene is 5.56×10^{-5} s^{-1} is calculated to be -68 °C by using the two values of Errede et al. [20] (such as the apparent first order rate constant of 9×10^{-6} s^{-1} at -78 °C and the apparent activation energy of the polymerization of 8.7 kcal/mol in the spontaneous polymerization of QM in toluene). The corresponding temperature for HCX is 30 °C. It is thus obvious that QM is much more reactive than HCX, i.e. HCX is a much more stable monomer than QM.

2-Cyano-7,7,8,8-tetrachloro-p-xylylene (CTCX)

2-Cyano-7,7,8,8-tetrachloro-p-xylene(CTCX) is prepared similarly by gas-phase dechlorination of 2,5-bistrichloromethylbenzonitrile on a copper mesh [59]. CTCX also readily undergoes spontaneous polymerization, the kinetics of which is approximately first order with respect to monomer at monomer concentrations higher than $2-3 \times 10^{-3}$ mol/l whereas it is second order below this monomer concentration. The apparent first-order rate constant has been determined as 1.3×10^{-4} s^{-1} at 30 °C at an initial monomer concentration of 1.0×10^{-2} mol/l. The apparent activation energy of the polymerization is 8.8 kcal/mol. In addition, the temperature at which the apparent rate constant is 5.6×10^{-5} s^{-1} has been found to be 15 °C for CTCX. The tendency of CTCX for homopolymerization is higher than that of HCX (30 °C) and much lower than that of QM (-68 °C).

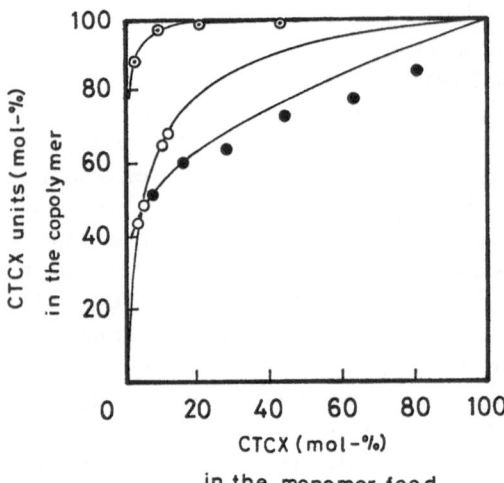

Fig. 5. Diagram of the composition of the copolymerization of CTCX with St. (○) refers to experimental values, (●) values of the copolymerization of HCX with St, and (⊙) values of the copolymerization of TCX with St. Solid lines are obtained by means of the theoretical equation using the monomer reactivity ratios (r_1(CTCX) = 12 and r_2(St) = 0.03 for this system, r_1(HCX) = 3.0 and r_2(St) = 0.02 at 50 °C for the HCX-St system, and r_1(TCX) = 85 and r_2(St) = 0 at 22 °C for the TCX-St system

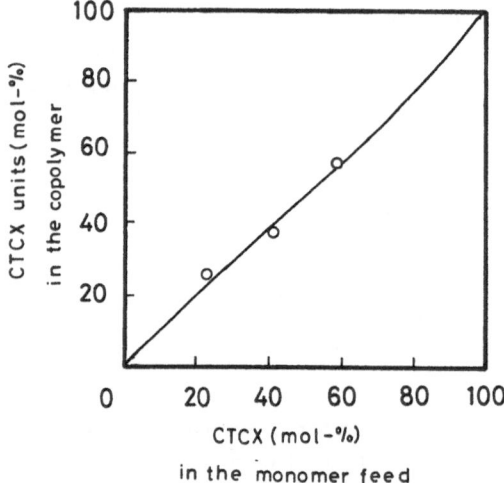

Fig. 6. Diagramm of the composition of the copolymerization of CTCX with HCX. (○) refers to experimental value, and solid line is calculate from the theoretical equation using r_1(CTCX) = 0.8 and r_2(HCX) = 0.95

In contrast to QM, three monomers of TCK, HCX and CTCX have been found to undergo spontaneous copolymerization with various vinyl monomers such as styrene (St), isoprene, vinyl acetate, acrylonitrile, and methyl methacrylate [58,59,60]

For the copolymerization systems of TCX-St, HCX-St, CTCX-St, (Fig. 5) CTCX-

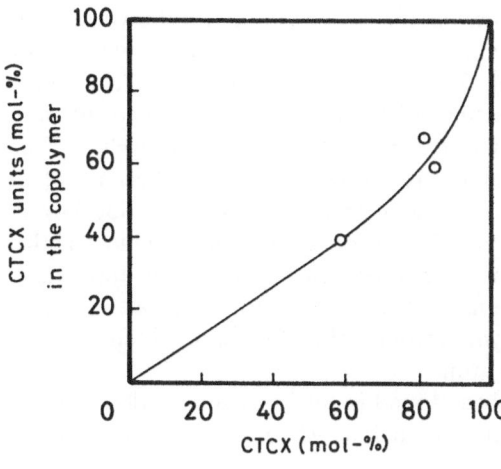

Fig. 7. Diagram of the composition of the copolymerization of CTCX with TCX. (○) refers to experimental value, and solid line is calculated from the theoretical equation using r_1 (CTCX) = 0.25 and r_2 (TCX) = 1.7

HCX, (Fig. 6) and CTCX-TCX (Fig. 7) the following monomer reactivity ratios have been obtained; r_1(TCX) = 85 and r_2(St) = 0 at 22 °C [60], r_2(HCX) = 3 ± 0.8 and r_2(St) = 0.02 ± 0.05 at 50 °C [58], r_1(CTCX) = 12 ± 6 and r_2(St) = 0.03 ± 0.02 at 20 °C [59], r_1(CTCX) = 0.8 and r_2(HCX) = 0.95 at 20 °C, and r_1(CTCX) = 0.25 and r_2(TCX) = 1.7 at 20 °C [59]. The relative reactivities of TCX, HCX, and CTCX toward the active site of the polymer chain with a terminal St unit have been estimated from comparison of the reciprocals of r_2(St) of the systems of TCX-St, HCX-St and CTCX-St: TCX(1/0) > HCX(1/0.02) ≧ CTCX(1/0.03). The monomer reactivity ratios are thereby assumed to remain essentially unchanged in the temperature range of the polymerization (20 to 50 °C). A comparison of the reciprocals of r_1(CTCX) of the systems of TCX-CTCX and HCX-CTCX gives another order of the

Table 2. π-Electron density, frontier density, and free valence at the exocyclic carbon of CTCX, HCX and TCX[a]

	π-Electron density		Frontier density	Free valence
	Ground state	Singlet Excited state		
CTCX[b] 1	0.9642	1.0169	0.4048	0.412
2	0.9556	1.0675	0.4598	0.421
HCX	0.9691	1.0676	0.4469	0.424
TCX	0.9869	1.0716	0.4620	0.440

[a] Calculated by the ASMO-SCF method;

relative reactivities of the monomers toward the active site of the polymer chain with a terminal CTCX unit: TCX (1/0.25) >HCX (1/0.8) ≧ CTCX (1/1). Both reactivity orders are in good agreement. In addition, the relative reactivities have been compared with some parameters estimated by quantum chemical calculations as π-electron density, frontier electron density, and free valence at exocyclic carbon (Table 2). CTCX has different values at the two exocyclic carbons because of its unsymmetric structure. The results are in agreement with the relationship of Kooyman and Farenhorst [61] between the relative reactivity of the trichloromethyl radical toward the aromatic hydrocarbons (such as benzene, naphthalene, anthracene, pyrene, etc.) and the highest free valence index of the hydrocarbon and also with the relationship of Hush [12] between the polymerizability of QM compounds and the free valence at their corresponding exocyclic atom.

Another interesting phenomenon has been found in the copolymerization of HCX with St starting with a high St monomer feed such as 92.3 mol%. A change of the

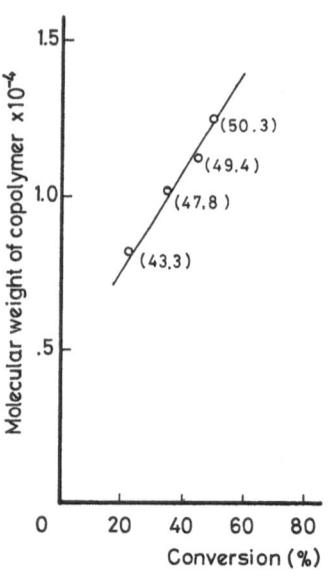

Fig. 8. Relationship between the molecular weight of the copolymer and conversion in the copolymerization of HCX with St starting with high St monomer concentration (92.3 mol%) Figures in brackets refer to the content of the St unit in mol% in the copolymer obtained

content of the St unit in the copolymers with conversion can be observed similarly as in common vinyl copolymerization but its magnitude is very small (see Fig. 8). The molecular weight of the obtained copolymers increases significantly with conversion as shown in Fig. 8. This implies that this polymerization partially proceeds by a stepwise addition mechanism, i.e. it exhibits a somewhat "living character" which is in sharp contrast to conventional free-radical vinyl polymerization. Therefore, it may be presumed that the active site of a polymer molecule with a terminal HCX unit reacts not only with St and HCX monomers via the radical addition mechanism but also with HCX via radical coupling.

Perchloro-p-xylylene prepared by Ballester et al. [62] exhibits no tendency for poly-merization and is very stable even at elevated temperature.

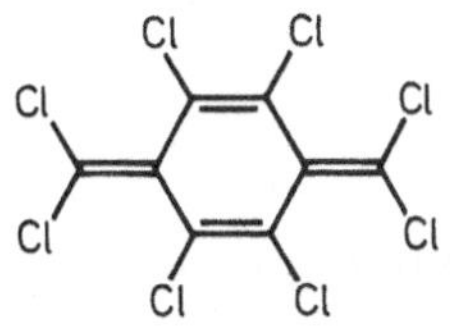

Perchloro-p-xylylene

6 Polymerization Behavior of Quinodimethanes as Acceptor Monomer

p-Benzoquinone (BQ) which displays electron-accepting properties is a well-known inhibitor [63,64] and retarder [65] in free-radical polymerization; it undergoes copoly-merization with styrene (St) despite its very low susceptibility to copolymerization [65,66]. p-Chloranil (CA), which is presumed to be a much stronger electron acceptor than BQ, undergoes alternating copolymerization with St in the presence of free-

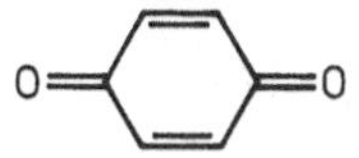

p-Benzoquinone (BQ)

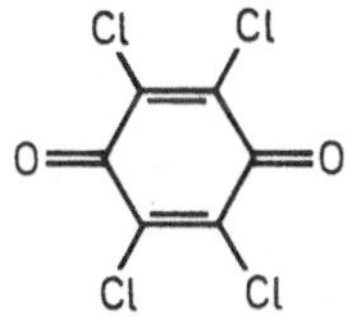

p-Chloranil (CA)

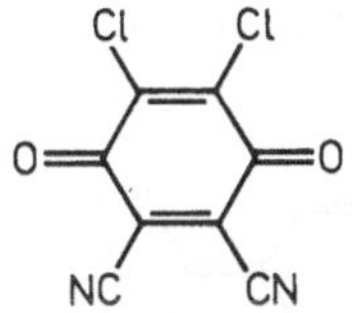

2,3-Dichloro-5,6-dicyano-p-benzoquinone (DDQ)

radical initiators [67,68]. 2,3-Dichloro-5,6-dicyano-p-benzoquinone (DDQ), an even stronger electron acceptor than CA, is also alternatingly copolymerized with St even in the absence of a free-radical initiator [69,70]. The relative reactivity of these benzoquinones as acceptor monomers toward the polymer radical with a terminal St unit is closely related to their electron-accepting character [70,71].

Chemists of du Pont described the preparation of a series of new compounds with electron-accepting properties such 7,7,8,8-tetracyanoquinodimethane (TCNQ) [72]

7,7,8,8-tetrakis(methoxycarbonyl)-quinodimethane (TMCQ) [72] 7,7,8,8-tetrakis(ethyl-sulfonyl)quinodimethane (TESQ) [73] and 11,11,12,12-tetracyanonaphtho-2,6-

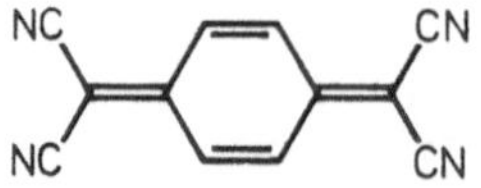

7,7,8,8 - Tetracyanoquinodimethane (TCNQ)

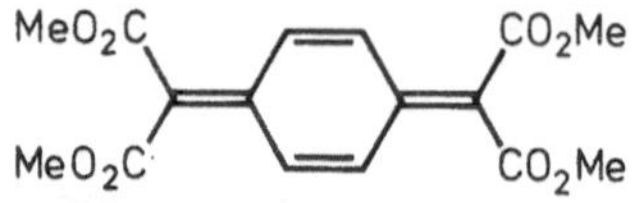

7,7,8,8 - Tetrakis (methoxycarbonyl) -quinodimethane (TMCQ)

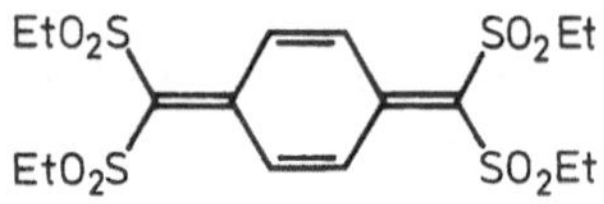

7,7,8,8 - Tetrakis (ethylsulfonyl) -quinodimethane (TESQ)

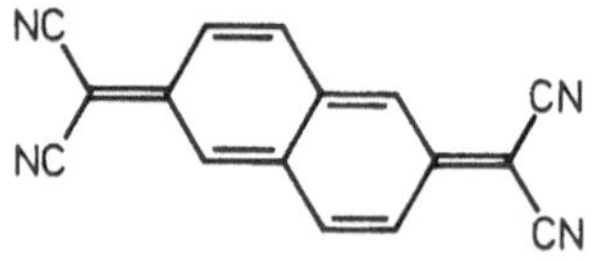

11,11,12,12 - Tetracyanonaphtho - 2,6 - quinodimethane (TNAP)

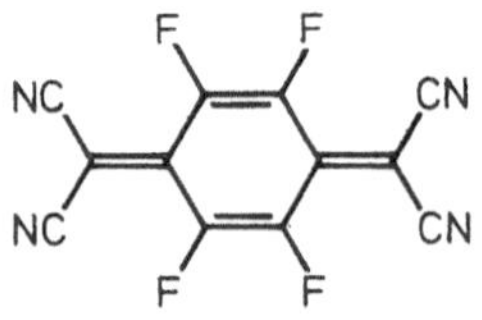

2,3,5,6 - Tetrafluoro -7,7,8,7 - tetracyanoquinodimethane (TCNQF$_4$)

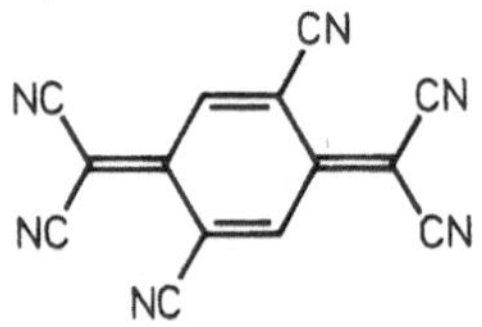

2,5,7,7,8,8 -Hexacyanoquinodimethane (TCNQ(CN)$_2$)

quinodimethane (TNAP) [74] in the eary 1960's. In addition, 2,3,5,6-tetrafluoro-7,7,8,8-tetracyanoquinodimethane (TCNQF$_4$) [75] and 2,5,7,7,8,8-hexacyanoquino-dimethane (TCNQ(CN)$_2$) [75], which display stronger electron-accepting properties than TCNQ, were prepared in 1975. Those compounds have been extensively studied due to their powerful electron-accepting character, e.g., on their charge-transfer complexes with high electric conductivity refered to as organic metal [76]. TCNQ was reported to initiate a cationic polymerization of alkyl vinyl ethers [77,78]. On the other hand, polymerization of these quinodimethanes as acceptor monomers had not been studied in detail until the spontaneous alternating copolymerization of TCNQ with St was reported in 1978 [79].

6.1 TCNQ-St System [79]

When TCNQ is mixed with a styrene solution in acetonitrile, a dark red color developes instantaneously. This color is explained by the formation of a charge-transfer complex between TCNQ and St. On standing at room temperature for a day, TCNQ dissolves slowly in acetonitrile and reacts with St at the interface of the TCNQ crystals, thereby producing a gelatinous shell of swollen pink-colored copolymer. This copolymer is insoluble in conventional organic solvents such as benzene, chloroform, and acetone, whereas in aprotic polar solvents such as N,N-dimethylformamide and dimethyl sulfoxide it swells at room temperature and eventually dissolves on prolonged heating at elevated temperatures (e.g. at 80 °C). Elemental analysis and NMR data reveal that the copolymer is a truely alternating copolymer. Its NMR spectrum only contains two kinds of peaks in the ranges δ 7.0–7.5 ppm and 3–3.5 ppm. The latter peak is assigned to the methine and methylene protons of the St units which are much more deshielded than the corresponding protons of homopolystyrene whose peaks generally appear between δ 1.0 and 2.0 ppm [64]. This deshielding is presumed to arise primarily from the powerful electron-withdrawing effect of the neighboring dicyanomethylene groups when the St unit directly links two TCNQ units on its both sides.

6.2 System of St with TCNQF$_4$ [81], TNAP [82], and TCNQ(CN)$_2$ [82]

TCNQF$_4$ displays considerably stronger electron-accepting properties [83] and is better soluble in organic solvents than TCNQ. Thus, it is conveniently used for kinetic studies. The electron-accepting character of TNAP is intermediate between that of TCNQ and TCNQF$_4$, and TCNQ(CN)$_2$ displaying the strongest electron-accepting properties [83]. The addition of St to a solution of TCNQ, TNAP, or TCNQF$_4$ in acetonitrile causes a deepening of the color of respective acceptor solution, corresponding to an absorption of the generated charge-transfer complexes. TCNQ-St, TNAP-St and TCNQ-$_4$-St systems absorb light in the range of 450 to 580 nm, 530 to 630 nm, and 500 to 750 nm, respectively [82]. The charge-transfer transition absorption of the TCNQ(CN)$_2$-St system cannot be measured because the absorbance of the mixture decreases so rapidly that it disappears completely within a minute, probably due to very rapid polymerization. TCNQF$_4$, TNAP, and TCNQ(CN)$_2$ undergo alternating copolymerization with St without any initiator, similarly as in the case of TCNQ. Kinetics studies of the spontaneous alternating copolymerization of TCNQF$_4$-St system revealed that the copolymerization follows the three-halves order kinetics with respect to each concentration of TCNQF$_4$ and St [81]. The copolymerizations of the TCNQ-St [81], and TNAP-St [82] systems were found to obey the same kinetics. On the other hand, the copolymerization of the TCNQ(CN)$_2$-St system follows first order kinetics both with respect to TCNQ(CN)$_2$ and St [82]. The three-halves order kinetics has previously been found for the spon-

Table 3. Rate constant, overall activation energies, and half-life periods of the copolymerization of the TCNQ-St, TNAP-St, TCNQF$_4$-St, and TCNQ(CN)$_2$-St Systems

Systems	Order	Rate constant[a] $kp \; l^{1/2} \cdot mol^{-1/2} \cdot s^{-1}$	Overall activation energy, KJ/mol	Half-life time, min
TCNQ-St	1.5	2.75×10^{-3}	72.3	1000[b]
TNAP-St	1.5	2.14×10^{-2}	68.0	151
TCNQF$_4$-St	1.5	5.29×10^{-1}	69.2	4.0
TCNQ(CN)$_2$-St	1.0	$1.05 \times 10^{-3}, s^{-1}$	69.7	1.2

[a] Values at 34.5 °C for the TCNQ-St, TNAP-St, and TCNQF$_4$-St systems and at 10 °C for the TCNQ(CN)$_2$-St system, respectively;
[b] Calculated from the rate constant

taneous alternating copolymerizations of the systems p-dioxene-maleic anhydride and 1,2-dimethoxyethylene-maleic anhydride [84]. A similar copolymerization reaction scheme [84] has been proposed: 1) donor and acceptor monomers form a charge-transfer complex and the monomolecular reaction of the complex gives the propagating radical species, 2) the radical species adds the complex to give an alternating copolymer, 3) termination takes place between the propagating polymer radicals which are assumed to be in stationary state. The first order kinetics observed in the copolymerization of the TCNQ(CN)$_2$-St system was also found for the TCNQ-methyl methacrylate (MMA) system as mentioned latter. The rate constants, overall activation energies and half-life periods (the time to decrease in a half concentration of the acceptor monomer under a given monomer concentration) of the systems, TCNQ-St, TNAP-St, TCNQF$_4$-St, and TCNQ(CN)$_2$-St, are compiled in Table 3. These systems have similar overall activation energies of copolymerization. The TCNQ(CN)$_2$-St system copolymerizes about 1000times as rapidly as the TCNQ-St system. The rates of copolymerizations are closely ralated to the electron-accepting character of acceptor monomer such as electron affinity (EA) (see Table 5).

When acceptor monomers with a low positive e value of the Alfrey-Price Q-e scheme such as MMA (e = 0.4) [85], methyl acrylate (MA) (e = 0.6) [85] and acrylonitrile (AN) (e = 1.2) [85] are used as comonomers in the copolymerization with TCNQF$_4$, it has been found [81] that MMA is alternatingly and spontaneously copolymerized, MA undergoes alternating copolymerization only by means of a radical initiator, and AN is not susceptible to copolymerization. It is noteworthy that MMA and MA with a low positive e value undergo alternating copolymerization as donor monomer instead of acceptor monomer with TCNQF$_4$ which is a very strong electron acceptor monomer. This alternating tendency in the TCNQF$_4$-MMA and TCNQF$_4$-MA systems cannot be explained in terms of the Alfrey-Price Q-e scheme because all monomers in this scheme have positive e values and repulsive forces instead of attractive forces would be expected. It has therefore been proposed that the great difference in the polar character between TCNQF$_4$ and the alternatingly copolymerizable comonomers, which causes a charge-transfer interaction, is one of the primary factors responsible for their alternating tendency. In addition, TCNQ, which is a weaker acceptor monomer than TCNQF$_4$, also copolymerizes alternatingly and spontaneously with MMA whereas MA is not copolymerizable with

TCNQ [86]. The rate of spontaneous alternating copolymerization between TCNQ and MMA is about one thousandth as slow as that between TCNQF$_4$ and MMA [86]. Moreover, the slow rate of the copolymerization obeys first-order kinetics with respect to TCNQ monomer concentration [86].

6.3 Modes of Copolymerization of Vinyloxy Monomers with Electron-Accepting Quinodimethanes

As vinyloxy monomers n-butyl vinyl ether (nBVE), isobutyl vinyl ether (iBVE), 2-chloroethyl vinyl ether (CEVE), phenyl vinyl ether (PhVE), and vinyl acetate (VAc) have been used. The electron-donating character of these compounds is compared by means of Taft [87] and Hammett [88,89,90] substituent constants in regard to vinyloxy and vinyl compounds as compiled in Table 4. Stille et al. [77,78] reported that TCNQ initiates the cationic homopolymerization of alkyl vinyl ethers in acetonitrile, indicating a powerful electron-accepting character of TCNQ which causes an electron transfer. In the polymerization of TCNQ with each of the five monomers in acetonitrile, it has been found [91] that the first two vinyl ethers homopolymerize whereas the last three monomers copolymerize in an alternating fashion with TCNQ. The two modes of polymerization are consistently correlated with the electron-donating character of the vinyloxy monomers, e.g. Taft substituent constants for vinyloxy compounds [91]. Moreover, when other electron-accepting quinodimethane derivatives such as TCNQ(CN)$_2$ [82] TCNQF$_4$ [81], DDQ [70], TNAP [82], TESQ [96] and TMCQ [97] are used, the modes of polymerization indicated in Table 5 are observed. These results except for the case of TESQ suggest that the modes of the polymerization are also correlated with the electron-accepting character of these monomers. It is concluded, therefore, that the difference in polar character between the donor and acceptor monomers strongly affects the determination of the mode of polymerization. This large difference is probably responsible for an electron-transfer reaction between donor and acceptor monomers.

Table 4. Taft's and Hammett's substituent constants of vinyloxy and vinyl compounds

Substituents R	Vinyloxy compounds			Substituents R'	Vinyl compounds		
	Taft's constant[87]	Hammett's constant			Hammett's constant		e value [85]
	σ^*	σ_m	σ_p		σ_m [88]	σ_p [88]	
CH$_3$CO	+1.65	+0.376 [88]	+0.502 [88]	CH$_3$COO	+0.39	+0.31	−0.220
C$_6$H$_5$	+0.600	+0.06 [88]	−0.01 [88]	C$_6$H$_5$O	+0.25	−0.32	−1.210
ClCH$_2$CH$_2$	+0.385		(+0.18 [90])				−1.410
C$_2$H$_5$	−0.100	−0.07 [88]	−0.151 [88]	C$_2$H$_5$O	+0.1	−0.24	−1.170
n-C$_3$H$_7$	−0.115	−0.05 [89]	−0.126 [89]	n-C$_3$H$_7$O	+0.1	−0.25	−1.520
iso-C$_4$H$_9$	−0.125		−0.115 [90]				−1.770
n-C$_4$H$_9$	−0.130	−0.07 [89]	−0.161 [89]	n-C$_4$H$_9$O	+0.1	−0.32	−1.200
iso-C$_3$H$_7$	−0.190	−0.07 [89]	−0.151 [88]	iso-C$_3$H$_7$O	+0.1	−0.45	−1.310
tert-C$_4$H$_7$	−0.300	−0.10 [88]	−0.197 [88]				−1.580

^{a} Value for ClCH$_2$

Table 5. Modes polymerization in a acetonitrole

	TCNQ(CN)$_2$ [82]	TCNQF$_4$ [81]	DDQ [70]	TNAP [82]	TCNQ [91]	TESQ [96]	TMCQ [97]
Electron Affinity (eV)		3.22 [92]	3.00 [92]		2.88 [92]		
Reduction Potential (V)	0.65 [83]	0.53 [83]	0.51 [93]	0.21 [74] 0.20 [83]	0.17 [83] −0.2 [95]	0.092 [94]	−0.83 [95]
VAC	Adduct	Alternating copolymer	Alternating copolymer	Adduct	Alternating copolymer	Adduct	Alternating copolymer
PhVE	Adduct	Alternating copolymer	Alternating copolymer	Adduct	Alternating copolymer	Adduct	Alternating copolymer
CEVE	homopolymer	homopolymer	homopolymer	homopolymer	Alternating copolymer	homopolymer	Alternating copolymer
nBVE	homopolymer	homopolymer	homopolymer	homopolymer	homopolymer	homopolymer	Alternating copolymer
iBVE	homopolymer	homopolymer	homopolymer	homopolymer	homopolymer	homopolymer	Alternating copolymer

Furthermore, it has been found [98] that the mode of polymerization of TCNQ with CEVE depends upon the solvent used. Thus, an alternating copolymer is obtained in acetonitrile, whereas in ethylene carbonate a homopolymer of CEVE results. Low molecular weight porducts composed of TCNQ and CEVE units are obtained when dimethylsulfoxide (Me$_2$SO) and N-methylformamide are employed as solvents. This difference in the mode of polymerization may be explained in terms of the polarity and basicity of the solvent. The dielectric constants of acetonitrile, Me$_2$SO, ethylene carbonate, and N-methylformamide are 37.5, 46.68, 89.6, and 182.4, respectively [99]. The frequency shifts, Δv_{OH}, of phenol with acetonitrile, ethylene carbonate, and Me$_2$SO are 155, 159, and 350 cm^{-1}, repsectively [100]. Frequency shifts, Δv_{OH}, of p-fluorophenol with N-methylformamide and Me$_2$SO are 271 and 367 cm^{-1}, respectively [101]. It is therefore obvious that the basicity of these solvents increases in the following order: acetonitrile $\leq$ ethylene carbonate $<$ N-methyl-formamide $<$ Me$_2$SO. From the difference in the dielectric constants the following order of reactivity of the electron transfer reaction between TCNQ and CEVE may be established: N-methylformamide $>$ ethylene carbonate $>$ Me$_2$SO $>$ acetonitrile. Since N-methylformamide is more basic than ethylene carbonate, the cationic end of the radical cation species formed by the electron transfer reaction can be more tightly solvated by solvent molecules, probably leading to a deactivation of the cationic end [102]. Another free radical end may add monomer molecules to give low molecular weight products composed of both monomer units. The formation of low molecular weight products in Me$_2$SO also may be attributed to its high basicity. The drastic difference in the mode of polymerization between acetonitrile and ethylene carbonate conceivably arises primarily from the polarity of these solvents. The more polar ethylene carbonate permits the electron transfer reaction between TCNQ and CEVE to occur, while the less polar acetonitrile does not. Because both solvents have a low basicity of similar magnitude, the cationic polymer end is only weakly solvated and does not inhibit the cationic polymerization.

6.4 Amphoteric Behavior of TMCQ [97] and TECQ [103] in Alternating Copolymerization

TMCQ and tetrakis(ethyoxycarbonyl)quinodimethane (TECQ) are quinodimethane derivatives with an electron-withdrawing functional group similar to that of TCNQ, but are much weaker electron acceptor than TCNQ. A study [103] of the charge-transfer absorption bands between TMCQ or TECQ and conventional donor compounds revealed that TMCQ and TECQ display electron-accepting properties which are however much weaker than those of TCNQ. TECQ has been found to exhibit only a slightly weaker electron-accepting nature than TMCQ. Moreover, it has been found that TMCQ [97] and TECQ [103] behave as electron donor to form charge-transfer complexes with TCNQ which has very strong electron-accepting properties. A comparison of the absorption bands reveals that TECQ [97] displays only a slightly

7, 7, 8, 8 - Tetrakis (ethoxycarbonyl) - quinodimethane (TECQ)

more electron-donating character toward TCNQ than TMCQ. Consequently, it is concluded [103] that TMCQ and TECQ exhibit an amphoteric polar character in the formation of their charge-transfer complexes. Their amphoteric polar nature may be explained consistently in terms of a π-electron density scheme. It is reasonable to speculate that St has the highest π-electron density followed by TECQ and TMCQ, the former exhibiting only a slightly higher density than the latter, and TCNQ clearly has the lowest density. The sufficiently large difference in the π-electron density gives therefore rise to the formation of charge-transfer complexes between St and TECQ or TMCQ as well as between TECQ or TMCQ and TCNQ. Moreover, it is understandable that TECQ is a weaker electron acceptor than TMCQ while TECQ is a stronger electron donor than TMCQ.

TMCQ [97] and TECQ [103] copolymerize alternatingly and spontaneously as acceptor monomers with such conventional electron-donating comonomers as St, iBVE,

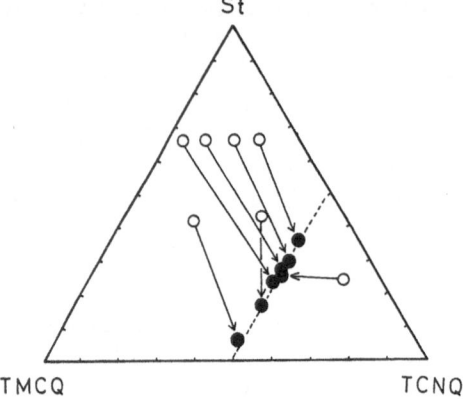

Fig. 9. Triangular diagram of the composition of the terpolymer of TCNQ, TMCQ, and St. (O): feed composition, (●) terpolymer composition. Arrows denote change in the composition from the feed to the terpolymer obtained

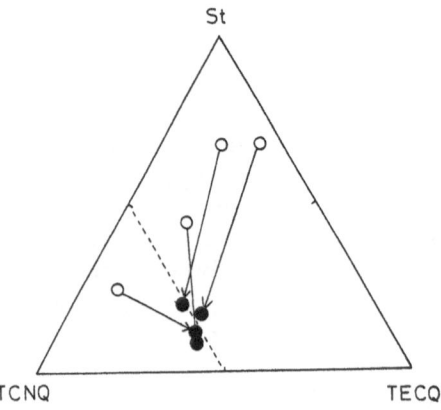

Fig. 10. Triangular diagram of the composition of the terpolymer of TECQ, TCNQ, and St: (O) feed composition, (●) terpolymer composition. Arrows denote change in the composition from the feed to the terpolymer obtained

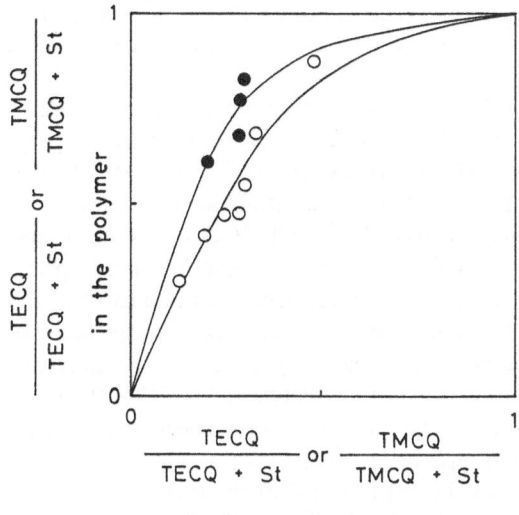

Fig. 11. Diagrams of the composition of the terpolymerizations of TECQ, TCNQ, and St, and of TMCQ, TCNQ, and St as binary copolymerizations between TECQ and St and between TMCQ and St, respectively. The lines are calculated using $r_1(K_1/K_2) = 15 \pm 10$ and $r_2(K_2/K_1) = 0.5 \pm 0.3$ for the terpolymerization of the TECQ-TCNQ-St system (●), and $r_1(K_1/K_2) = 7 \pm 3$ and $r_2(K_2/K_1) = 0.7 \pm 0.3$ for the terpolymerization of the TMCQ-TCNQ-St systrm (○), respectively

nBVE, CEVE, PhVE, and VAc. TMCQ and TECQ are also copolymerized alternatingly and spontaneously as donor monomers with a very poweful acceptor monomer such as TCNQ. This amphoteric behavior in alternating copolymerization was found first in the terpolymerization [97] of St, TMCQ and TCNQ, the obtained terpolymers containing 50 mol% of TCNQ units regardless of the monomer feed ratio. This indicates that TMCQ and St copolymerize as donor monomers with TCNQ which is a strong acceptor, contrary to the expectation that TMCQ would copolymerize as an acceptor monomer. The compositional relationships of the terpolymerization of St, TMCQ and TCNQ, as well as of St, TECQ, and TCNQ are shown in the diagrams of Figs. 9 and 10, respectively, where open and closed circles refer to the monomer feed and the terpolymer composition, respectively. The terpolymerization composition relationships can be illustrated by the composition diagrams of the binary copolymerization between TMCQ and St and between TECQ and St (Fig. 11), because the content of the TCNQ unit is always constant (50 mol%) in any run. According to the mechanism of complex formation [104] in the alternating copolymerization, the apparent monomer reactivity ratios of the complexes are calculated as follows: $r_1(K_1/K_2)$ (TMCQ-TCNQ complex) = 7 ± 3 and $r_2(K_2/K_1)$ (St-TCNQ complex) = 0.7 ± 0.3 for the St-TMCQ-TCNQ system and $r_1(K_1/K_2)$ (TECO-TCNQ complex) = 15 ± 10 and $r_2(K_2/K_1)$ (St-TCNQ complex) = 0.5 ± 0.3 for the St-TECQ-TCNQ system. The relative reactivities of the TMCQ-TCNQ and TECQ-TCNQ complexes toward the polymer radical with a terminal St-TCNQ complex unit are as follows: St-TCNQ complex (1) < TMCQ-TCNQ complex (1.4) < TECQ-TCNQ complex (2.0). Thus, the TECQ-TCNQ complex is more reactive than the TMCQ-TCNQ complex. The reactivity of these complexes coincides with their electron-donating efficiency with respect to TCNQ, TECQ being a more electron donor than TMCQ.

The terpolymerization of TECQ-TMCQ-St [103] using monomer feed ratios (mol) of TECQ/TMCQ/St = 14.2/14.9/70.9 and 11.2/36.4/52.4 at 60 °C gave the terpolymers (conversion of 11.5 and 7.6%) with the ratios (mol) TECQ/TMCQ/St = 22.6/27.4/50.0 and 10.2/39.8/50.0, respectively. From the difference of the ratios of the TECQ to TMCQ content in monomer feed (TECQ/TMCQ = 0.49/0.51 and 0.24/0.76) and in the terpolymers (TECQ/TMCQ = 0.45/0.55 and 0.20/0.80, respectively) it may be assumed that TMCQ is somewhat more reactive than TECQ in the alternating copolymerization with St, i.e. the TMCQ-St complex is more reactive than the TECQ-St complex. Consequently, this reactivity order of the acceptor monomers in their alternating copolymerization with St between TMCQ and TECQ is in good agreement with the electron-accepting character with respect to St in their charge-transfer complex formation. The amphoteric behavior of TECQ and TMCQ in the alternating copolymerizations are explained in terms of the π-electron density scheme proposed for explaining the same behavior in the formation of the corresponding charge-transfer complexes. However, it is difficult to speculate that the difference in the polar character between TMCQ and TECQ may be attributed to an inductive substituent effect such as the Hammett constants [88,105] and to a steric substituent effect such as the Taft constant [87] between methyl and ethyl groups because of the very small differences.

When TMCQ is heated above 175 °C or exposed to light, it polymerizes even though the product seems to be dimeric or trimeric [72]. Recently, Hall and Bentley [95] reported that TMCQ polymerizes with free radical and anionic initiators to give homopolymer with a melting point of about 300 °C and an intrinsic solution viscosity of 0.91 dl/g (as poly-carboxonium-salt). Thus, TMCQ readily undergoes homopolymerization. However, TECQ cannot be homopolymerized by means of AIBN, n-butyllithium, and boron trifluoride etherate [103]. Only when TECQ is kept in the crystal line state at room temperature for a month long, a white powder with molecular weight of 2600 ($\overline{DP}$ = 6.6), is formed in poor yield [103]. This product is insoluble in benzene. Therefore, it is obvious that TECQ exhibits only a very slight tendency to homopolymerize, in contrast to TMCQ. Moreover, the difference in homopolymerizability cannot be attributed to a steric substituent effect because of very small difference in the Taft constant [87] between methyl (Es = 0.00) and ethyl group (Es = −0.07). Presumably, the specific structure of the tetrakis(alkoxycarbonyl)-quinodimethane may amplify significantly the small difference between methyl and ethyl groups, resulting in a different amphoteric character of the monomers in the charge-transfer complexation as well as in alternating copolymerization and a difference in the homopolymerizability between TMCQ and TECQ.

6.5 Polymerization Behavior of TESQ [96]

Since the ethylsulfonyl group (δp = 0.68) exhibits the same electron-withdrawing power as cyano group (δp = 0.66) in the Hammett substituent constant [88,105], TESQ was expected to display a similar polymerization behavior as an acceptor monomer to TCNQ. However, it has been found that in the charge-transfer transition [96] TESQ (EA = 1.17 eV) exhibits a much lower electron affinity than TCNQ (EA = 2.84 eV) [92]. TESQ and St have been subjected to a spontaneous alternating copolymerization in nitromethane. However, when p-dioxane or dichloromethane is used instead of nitromethane, an alternating copolymer is not obtained, and the content of the St units is higher than 50 mol%. From this it may be assumed that cationic polymerization of St takes place simultaneously. The effect of the solvent cannot be explained in terms of its polarity and basicity [102]. 1-Phenylethanol and TESQ may readily undergo the dehydration to give poly-St, suggesting that TESQ and its hydrogenation product may initiate simultaneously both cationic polymerization of St and the alternating copolymerization of St with TESQ. Moreover, when the TESQ fraction in monomer feed is above 40 mol%, no copolymer but only 1:1 adduct in high yield is obtained, in contrast to the copolymerization of TCNQ and St.

TESQ initiates the cationic oligomerization of iBVE, nBVE, CEVE, and PhVE [103]. A reaction of TESQ with VAc has not been observed. Therefore, TESQ is considered to be more acidic than TCNQ from the observed modes of polymerization of a series of those vinyloxy monomers whereas TESQ exhibits a lower electron affinity than TCNQ. Concerning these differences between TESQ and TCNQ, it should be taken into account that in the π-conjugation in which the substituents and the quinodimethane part are involved, the 3p orbital of sulfur participates for the ethylsulfonyl group, and 2p orbital of carbon for the cyano group. Price and Oae [106] suggested that the 2p–3p π-bond is less stable than the 2p–2p π-bond. According to the theory of hard and soft acids and bases it may be assumed that TESQ is a much harder acid than TCNQ.

6.6 7,8-Dialkoxycarbonyl-7,8-Dicyanoquinodimethanes [107,108]

These compounds (with methoxyl and ethoxyl group) were prepared independently by Hall et. al. [108] and Iwatsuki et. al. [107]. They have been expected to exhibit properties which are intermediate between TCNQ and tetrakis(alkoxycarbonyl)-

7,8 - Dialkoxycarbonyl -7, 8 - dicyanoquinodimethane

R = CH₃ (MCQ)

R = C₂H₅ (ECQ)

R = C₄H₉ (BCQ)

quinodimethane, e.g. TMCQ [95,97] and TECQ [103] because of their chemical structure. These electron-accepting character has been found to the consistent with expectations [107,108]. On the other hand, they copolymerize with p-chlorostyrene or St in a conventional random fashion [109], while TCNQ [79] or TMCQ [95,97] and St undergo alternating copolymerization. Moreover, they may readily be homopolymerized in the presence of radical and anionic initiators to high molecular weight polymer [107,108] (Table 6). When triethylamine is used as the anionic initiator, BCQ polymerizes almost as in a living anionic polymerization to give the polymer with a molecular weight over a million [109]. These monomers can undergo an anionic polymerization in presence of tetrahydrofuran, dimethyl sulfoxide, N,N-dimethyl-

Table 6. Homopolymerization of 7,8-dicyano-7,8-dimethoxycarbonylquinodimethane MCQ in dichloromethane[a]

Run no	2a mg	Catalyst		Time, hr	Temp., °C	Conv., %	^{n}sp/c[b] dl · g^{-1}
1	100.1	—	—	3.0	60	38.8	0.46
2	149.9	AIBN[c]	0.006 mmol	0.5	60	90.0	0.26
3	99.6	BF$_3$ · OEt$_2$	0.1 ml	21.8	0	0	—
4	96.3	n-BuLi[d]	one drop	0.5	0	82.5	0.10
5	80.2	Pyridine	0.011 mmol	0.5	0	54.0	0.21
6	100.0	Et$_3$N	0.006 mmol	0.5	0	100	0.37
7	99.9	Pyrrolidine	0.016 mmol	0.5	0	100	0.34

[a] Solvent: 20 ml. precipitant: methanol. nitrogen, atmosphere;
[b] Solvent: concentrated H$_2$SO$_4$. temp. 40 °C;
[c] AIBN: 2,2'-azobisisobutyronitrile;
[d] n-Butyllithium solution in benzene

formamide, methanol, acetonitrile etc., whereas when water or acetic acid is added, their polymerization is effectively inhibited [109]. The various properties of this polymer will be described in the near future. Nevertheless, it is obvious that unsymmetrically substituted quinodimethanes, bearing different substitutents at the exocyclic carbons, exhibit a very interesting polymerization behavior as electrophilic monomers.

7 References

Reviews
   i) Errede, L. A., Szwarc, M.: "Chemistry of p-xylylene, its analogous and polymers" Quart. Rev. (London) *12* 301 (1958)
  ii) Minoura, Y.: "Polymerization of p-xylylene in new polymerization reactions" (ed.) Saegusa, T. p. 7, Kyoto, Kagaku-Dojin Co. (1971) (in Japanese)
 iii) Iwatsuki, S.: "Poly-p-xylylenes", Kobunshi *23* 135 (1974) (in Japanese)
  1. Szwarc, M.: Nature *160* 403 (1947)
  2. Szwarc, M.: Disc. Faraday Soc. *2* 46 (1947)
  3. Szwarc, M.: J. Chem. Phys. *16* 128 (1948): J. Polymer Sci. *6* 319 (1951)
  4. Corey, R. S., Haas, H. C., Kane, M. W., Livingtongston, D. I.: J. Polym. Sci. *13* 137 (1954)
  5. Kaufman, M. H., Mark, H. F., Mesrobian, R. B.: J. Polym. Sci. *13* 3 (1954)
  6. Auspos, L. A., Hall, L. A. R., Hubbard, J. K., Kirk, Jr., W. Schaefgen, J. R., Speck, S. B.: J. Polym. Sci. *15* 9, 19 (1955)
  7. Schaefgen, J. R.: J. Polym. Sci. *15* 203 (1955)
  8. Farthing, A. C.: J. Chem. Soc. *1953* 3261
  9. Brown, C. J., Farthing, A. C.: J. Chem. Soc. *1953* 3270
 10. Coulson, C. A., Craig, D. P., Maccoll, A., Pullman, M. A.: Disc. Faraday Soc. *2* 36 (1947)
 11. Namiot, A. I., Dyatkin, M. E., Syskin, Y. K.: Compt. Rend. Acad. Sci. USSR *48* 267–9 (1945); C. A. *40* 4927 (1946)
 12. Hush, N. S.: J. Polym. Sci. *11* 289 (1953)
 13. Coppinger, G. M., Bauer, R. H.: J. Phys. Chem. *67* 2846 (1963)
 14. Evans, D. F.: J. Chem. Soc. *1959* 2753
 15. Errede, L. A., Landrum, B. F.: J. Am. Chem. Soc. *79* 4952 (1957)
 16. Errede, L. A., Hopwood, Jr. S. L.: J. Am. Chem. Soc. *79* 6507 (1957).

17. Errede, L. A., Hoyt, J. M.: J. Am. Chem. Soc. 82 436 (1960)
18. Errede, L. A., Cassidy, J. P.: J. Org. Chem. 24 1890 (1959)
19. Errede, L. A., Cassidy, J. P.: J. Am. Chem. Soc. 82 3653 (1960)
20. Errede, L. A., Gregorian, R. S., Hoyt, J. M.: J. Am. Chem. Soc., 82 5218 (1960)
21. Errede, L. A., Hoyt, J. M., Gregorian, R. S.: J. Am. Chem. Soc. 82 5224 (1960)
22. Errede, L. A.: J. Am. Chem. Soc. 83 949 (1961)
23. Errede, L. A., Pearson, W. A.: J. Am. Chem. Soc. 83 954 (1961)
24. Errede, L. A.: J. Polym. Sci. 49 253 (1961)
25. Errede, L. A.: J. Am. Chem. Soc. 83 959 (1961)
26. Jacobson, R. A.: J. Am. Chem. Soc. 54 1513 (1932)
27. Vanscheidt, A. A., Mel'Nikova, E. P., Krakovyak, M. G., Kukhareva, L. A., Gladkovkii, G. A.: J. Polym. Sci. 52, 179 (1961)
28. Sisido, K., Kusano, N.: J. Polym. Sci. A1 2101 (1963)
29. Minoura, Y., Shiina, O., Okabe, S.: Kogyo Kagaku Zasshi 70 1243, 1247 (1967)
30. Lunk, H. E., Youngman, E. A.: J. Polym. Sci. A3 2983 (1965)
31. Gilch, H. G.: J. Polym. Sci. A-1, 4 1351 (1966)
32. Hoeg, D. F., Luck, D. I., Goldberg, E. P.: J. Polym. Sci. B 2 697 (1964)
33. Dunnavant, W. R., Markle, R. A.: J. Polym. Sci. A-1 3 3649 (1965)
34. Gilch, H. G., Wheelwright, W. L.: J. Polym. Sci. A-1 4 1337 (1966)
35. Wade, R. H.: J. Polym. Sci. B 5 567 (1967)
36. Fujita, T., Yoshioka, T., Soma, N.: J. Polym. Sci. Polym. Lett. Ed. 16 1515 (1978); J. Polym. Sci. Polym. Chem. Ed. 18 3253 (1980)
37. Korshak, V. V., Kelenikov, G. S., Kharchevikova, A. V.: Dokl, Akad. Nauk. USSR 56 169 (1947)
38. Shishido, K., Kato, S.: Kogyo Kagaku Zasshi 43 565, 952 (1940)
39. Winberg, N. E., Fawcett, F. S., Mochel, W. E., Theobald, C. W.: J. Am. Chem. Soc. 82 1428 (1960)
40. Taylor, L. D., Kolesinski, H. S.: J. Polym. Sci. B 1 117 (1963)
41. Ito, Y., Miyata, S., Nakatsuka, M., Saegusa, T.: J. Org. Chem. 46 1043 (1981)
42. Korshak, V. V., Sergeev, V. A., Shifikov, V. K., Burenko, P. S.: Vysokomol. Soed. 5 1957 (1963): C. A. 60 5643 (1964)
43. Gorham, W. F.: J. Polym. Sci., 4 3027 (1966)
44. Gorham, W. F.: ACS Polym. Prepr. 73 (1965)
45. Gorham, W. F.: Brit. Pat. 883937–883941 (1961); Ger. Pat. 1085673 (1960); C. A. 55 22920 (1961)
46. Brown, C. J., Farthing, A. C.: Nature 164 915 (1949)
47. Brown, C. J.: J. Chem. Soc. 1953 3265
48. Cram, D. J., Steinberg, H.: J. Am. Chem. Soc. 73 5691 (1951)
49. Cram, D. J., Cram, J. M.: Accounts Chem. Res. 4 204 (1971)
50. Vögtle, F., Neumann, P.: Angew. Chem. Intern. Ed. 11 73 (1972)
51. Pollart, D. F.: Am. Chem. Soc. Div. Pertol. Chem., Prepr. 10 175 (1965); C. A. 67 21512 (1967)
52. Pollart, D. F.: US Pat. 3149175, Ger. Pat. 1155444 (1963); C. A. 60 2817 (1964) US. Pat. 3247274 (1966); C. A. 65 8816 (1966)
53. Neigisch, W. D.: J. Polym. Sci. Polym. Lett. Ed. 4 531 (1966)
54. Chow, S. W., Loeb, W. E., White, C. E.: J. Polym. Sci. 13 2325 (1969)
55. Chow, S. W., Pilato, L. A., Wheelwright, W. L.: J. Org. Chem., 35 20 (1970)
56. Glich, H. G.: Angew. Chem. 77 592 (1965)
57. Glich, H. G.: J. Polym. Sci. A-1 4 438 (1966)
58. Iwatsuki, S., Kamiya, H.: Macromolecules 7 732 (1974)
59. Iwatsuki, S., Innoue, K.: Macromolecules 10 58 (1977)
60. Iwatsuki, S., Kokubo, T.: unpublished results
61. Kooyman, E. C., Farenhorst, E.: Trans, Faraday Soc. 49 58 (1953)
62. Ballester, M., Castaner, J., Riera, J.: J. Am. Chem. Soc. 88 957 (1966)
63. Bevington, J. C., Ghanem, N. A., Melville. H. W.: Trans. Faraday Soc. 51 946 (1955)
64. Tüdös, F.: J. Polym. Sci. 30 343 (1958)
65. Bevington, J. C., Ghanem, N. A., Melville, H. W.: J. Chem. Soc. 1955 2822

66. Bamford, C. H., Barb, W. C., Jenkins, A. D., Onyon, P. F.: "The Kinetics of Vinyl Polymerization by Radical Mechanism" p 188, 247 London Butterworth 1958
67. Breitenbach, J. W.: Can. J. Res. *28B* 507 (1950)
68. Hauser, C. F., Zutty, N. L.: J. Polym. Sci. A-1 *8* 1385 (1970)
69. Hauser, C. F., Zutty, N. L.: Macromolecules *4* 478 (1971)
70. Iwatsuki, S., Itoh, T.: J. Polym. Sci., Polym. Chem. Ed. *18* 2971 (1980)
71. Iwatsuki, S., Itoh, T.: Makromol. Chem. *182* 2161 (1981)
72. Acker, D. S., Hertler, W. R.: J. Am. Chem. Soc. *84* 3370 (1962)
73. Hertler, W. R., Benson, R. E.: J. Am. Chem. Soc. *84* 3475 (1962)
74. Diekmann, J., Hertler, W. R., Benson, R. E.: J. Org. Chem. *28* 2719, (1963)
75. Wheland, R. C., Martin, E. L.: J. Org. Chem. *40* 3101 (1975)
76. For instance, Torrance, J. B.: Accounts Chem. Res. *12* 79 (1979)
77. Aoki, S., Stille, J. K.: Macromolecules *3* 473 (1970)
78. Tarvin, R. F., Aoki, S., Stille, J. K.: Macromolecules *5* 663 (1972)
79. Iwatsuki, S., Itoh, T., Horiuchi, K.: Macromolecules *11* 497 (1978)
80. Bovey, F. A., Tiers, G. V. D., Filipovich, G.: J. Polym. Sci. *38* 73 (1959)
81. Iwatsuki, S., Itoh, T.: Macromolecules *15* 347 (1982)
82. Iwatsuki, S., Itoh, T., Saito, H., Okada, J.: Macromolecules, in press
83. Wheland, R. C., Gillson, J. L.: J. Am. Chem. Soc. *98* 3916 (1976)
84. Kokubo, T., Iwatsuki, S., Yamashita, Y.: Makromol. Chem. *123* 256 (1969)
85. Young, L. J.: "Polymer Handbook", (ed.) Brandrup, J., Immergut, E. H. vol. II, p. 387, New York Wiley-Interscience; 1975
86. Iwatsuki, S., Itoh, T.: Macromolecules *16* 332 (1983)
87. Taft, Jr., R. W.: "Steric Effects in Organic Chemistry" (ed.) Newman, M. S., Chapter 13, New York, Wiley 1968
88. McDaniel, D. H., Brown, H. C.: J. Org. Chem. *23* 420 (1958)
89. Charton, M.: J. Org. Chem. *28* 3121 (1963)
90. Jaffe, H. H.: Chem. Rev. *53* 191 (1953)
91. Iwatsuki, S., Itoh, T.: Macromolecules *12* 208 (1979)
92. Chen, E. C. M., Wentworth, W. E.: J. Chem. Phys. *63* 3183 (1975)
93. Peover, M. E.: J. Chem. Soc. *1962* 4540
94. Melby, L. R., Harder, R. J., Hertler, W. R., Mahler, W., Benson, R. E., Mochel, W. E.: J. Am. Chem. Soc. *84* 3374 (1962)
95. Hall, Jr., H. K., Bentley, J. H.: Polymer Bull. *3* 203 (1980)
96. Iwatsuki, S., Itoh, T., Shimizu, Y.: Macromolecules *16* 532 (1983)
97. Iwatsuki, S., Itoh, T.: Macromolecules *13* 983 (1980)
98. Iwatsuki, S., Itoh, T., Sadaike, S.: Macromolecules *14* 1608 (1981)
99. Riddick, J. A., Bunger, W. B.: Organic Solvents 3rd edit. p. 536 New York, Wiley 1970
100. Filgueiras, C. A. L., Huheey, J. E.: J. Org. Chem. *41* 49 (1946)
101. Arnett, E. M., Joris, L., Mitchell, E., Murty, T. S. S. R., Gorrie, T. M., Schleyer, P. V. R.: J. Am. Chem. Soc. *92* 2365 (1970)
102. Shirota, Y., Mikawa, H.: J. Macromol. Sci. Rev. Macromol. Chem. *C16(2)* 129 (1977–1978)
103. Iwatsuki, S., Itoh, T., Yokotani, I.: Macromolecules in press
104. Iwatsuki, S., Yamashita, Y.: Makromol. Chem. *104* 263 (1967)
105. Gordon, A. J., Rord, R. D.: "The Chemist's Companion" p. 146, New York, Wiley, 1972
106. Price, C. C., Oae, S.: "Sulfur Bonding" p. 5 New York Ronald Press Co., 1962
107. Iwatsuki, S., Itoh, T., Nishihara, K., Furuhashi, H.: Chem. Lett. *1982* 517
108. Hall, Jr. H. K., Cramer, R. J., Mulvaney, J. E.: Polym. Bull. *1982* 165
109. Iwatsuki, S., Itoh, T., Iwai, T.: unpublished results

Received July 13, 1983
T. Saegusa (Editor)

Author Index Volumes 1–58

Subject Index

M. Szwarc

Living Polymers and Mechanisms of Anionic Polymerization

1983. 69 figures. V, 187 pages. (Advances in Polymer Science, Volume 49). ISBN 3-540-12047-5

Contents: Introduction. – Thermodynamics of Polymerization. – Initiation of Anionic Polymerization. – Propagation of Anionic Polymerization. – Concluding Remarks. – References.

Michael Szwarc is the "father" of the term "living polymers" which he proposed for those macromolecules that may spontaneously resume their growth whenever fresh monomer is supplied to the system. While the conventional polymerization scheme created the impression that a terminationless polymerization is highly improbable, Szwarc and his associates demonstrated the terminationless character of anionic polymerization of vinyl monomers in the absence of impurities around 1956. Living polymers, although not named in this way, were described earlier by Ziegler. However, Szwarc was able to reveal their characteristic features: they do not die but remain acitve waiting for the next monomer. If the monomer added is different from the one previously used, a "block polymer" results. This, indeed, is the most versatile technique for synthesizing block polymers.
In this review, the author describes the mechanisms of anionic polymerization and pays special attention to the concept and application of living polymers. The role of various species, e.g. free ions, ion-pairs, triple ions etc. is stressed and their meaning is clarified. Their nature and the pertinent interrelations, both thermodynamic and kinetic, are explained. The article provides a background to anybody interested in organic ion reactions and ionic polymerization. (487 ref.)

Springer-Verlag
Berlin
Heidelberg
New York
Tokyo

Anionic Polymerization

With contributions by **L.J. Fetters, J. Lustoň,
R.P. Quirk, F. Vašš, R.N. Young**

1983. 21 figures, 30 tables. Approx. 200 pages.
(Advances in Polymer Science, Volume 56).
ISBN 3-540-12792-5

R.N. Young, R.P. Quirk, L.J. Fetters, **Anionic Polymerizations of Non-Polar Monomers Involving Lithium**
Carbanionic polymerizations involving lithium and non-polar monomers have achieved a position of special interest and importance as a result of the potential for obtaining systems lacking spontaneous termination reactions. The non-terminating nature of these systems facilitates kinetic studies, the preparation of polymers of narrow molecular-weight distributions and predictable molecular weights, the synthesis of block copolymers of uniform composition and molecular weight, and allows controlled termination reactions where star-shaped or comb-type polymers can be formed as well as chains having functional groups at one or both ends. – Thus, even though limited to relatively few monomers, anionic polymerizations have attracted both academic and commercial interest.

J. Lustoň, F. Vašš, **Anionic Copolymerization of Cyclic Ethers with Cyclic Anhydrides**
This article summarizes and analyzes the results obtained for anionic copolymerization of cyclic ethers with cyclic anhydrides. This reaction is of great practical importance especially as curing reaction of epoxy resins and is also used for preparation of linear polyesters with special functional pendant groups. – Emphasis is placed mainly on kinetics and mechanisms of copolymerization, the influence of epoxide and anhydride structure on copolymerization, the effect of type and structure of the initiator used, and on the course of copolymerization. The probable mechanisms are discussed. The copolymerization in the presence of proton-donor compounds is also considered as well as the effect of proton-donors. Data and theoretical views on non-catalyzed copolymerizations are included.

Springer-Verlag
Berlin
Heidelberg
New York
Tokyo